CW01338047

A BRIEF GUIDE TO ALIENS

ALSO BY ADAM FRANK

LIGHT OF THE STARS

ASTRONOMY: AT PLAY IN THE COSMOS

THE CONSTANT FIRE

ABOUT TIME

A BRIEF GUIDE TO ALIENS

From Myth to Modern Science

Adam Frank

HERO, AN IMPRINT OF LEGEND TIMES GROUP LTD
51 Gower Street
London WC1E 6HJ
United Kingdom
www.hero-press.com

First published in the USA by HarperCollins in 2023 under the title
The Little Book of Aliens
First published in the UK in this edition by Hero in 2024

© Adam Frank, 2024

The right of the author to be identified as the author of this work has been asserted in accordance with the Copyright, Designs and Patents Act 1988. British Library Cataloguing in Publication Data available.

Printed by Akcent Media, 5 The Quay, St Ives, Cambs, PE27 5AR

ISBN: 978-1-91505-413-5

All rights reserved. No part of this publication may be reproduced, stored in or introduced into a retrieval system, or transmitted, in any form or by any means (electronic, mechanical, photocopying, recording or otherwise), without the prior written permission of the publisher. This book is sold subject to the condition that it shall not be resold, lent, hired out or otherwise circulated without the express prior consent of the publisher.

TO PROF BRUCE BALICK.

I always tell incoming students that choosing a good PhD advisor is the most important decision they'll make in graduate school. I do this because I was so lucky in finding you. Thank you for your lived example of what a life in science should look like: creativity, kindness, delight, and precision.

CONTENTS

Introduction — xiii

CHAPTER 1
How Did We Get Here?
HOW OUR ANCIENT QUESTIONS ABOUT
ALIENS TOOK THEIR MODERN FORM — 1

A Really Old Question — 2
Alien Debates through History

Fermi's Paradox — 6
Is There a Great Silence?

The Drake Equation — 12
Asking the Right Questions

The UFOs Arrive — 19
*Kenneth Arnold Sees Saucers. Roswell Gets Busy.
The Government Reports*

Invasion of the Pop-Culture Extraterrestrials — 31
They're Here!

CHAPTER 2
So How Do We Do This?
CRITICAL IDEAS THAT SHAPED, AND STILL SHAPE,
OUR SEARCH FOR ALIENS — 35

Project Ozma — 38
The First Search

Habitable Zones — 43
Goldilocks in Orbit

Dyson Spheres — 47
Aliens Go Big with Megastructures

The Kardashev Scale — 51
How to Measure an Alien Civilization

CHAPTER 3
WTF UFOs and UAPs?
HOW THEY DO, OR DO NOT, FIT INTO THE SEARCH FOR ALIENS   56

The Giggle Factor   58
How Politics and UFOs Almost Killed the Search for Alien Life

Hoaxes and Hoaxers   61
A Good Con Never Dies

The McDonald Critique   64
So, About Those Unexplained Cases . . .

UFOs Become UAPs   67
The Modern Era Begins

How to Get Real About Ufos   74
What a True Scientific Study Would Look Like

CHAPTER 4
What If They Are Aliens?
IF UFOS ARE ET, HOW'D THEY GET HERE, AND WHAT THE HELL ARE THEY DOING?   78

Interstellar Travel   79
If Ufos Were Aliens, How Did They Get Here?

Alien Technology   86
Inside Luke Skywalker's Garage

Interdimensional Aliens   90
Hey, Man, Get off My Plane

But What Are They Doing Here?   96
The High-Beam Argument and Some Other Questions

CHAPTER 5
Cosmic Curb Appeal?
WHERE TO LOOK FOR ALIENS   99

The Origin of Life   100
The Miller-Urey Experiment and Abiogenesis

The Ocean Moons   106
    Who Knew?

Exoplanets   108
    The Revolution Will Be Telescoped

Planets Gone Wild   112
    The Super-Earth Enigma

Snowball Worlds and Ocean Worlds   115
    Winter Is Coming, and so Is the Flood

Ten Billion Trillion Chances to Roll the Dice   122
    The Pessimism Line and What It Tells Us

CHAPTER 6
The Cosmic Stakeout
HOW WE'RE GOING TO SPY ON ET   127

Biosignatures   129
    How to Find Life From a Distance

Technospheres And Noospheres   134
    When Smart Life Goes Boss

Technosignatures   137
    The Day the Earth Stood Still-Ish

Attack of the Alien Megastructures   140
    Boyajian's Star

Pollution, City Lights, and Glint   143
    What Alien Skies Can Tell Us about Aliens

Solar System Artifacts   148
    Did You Leave These?

Was 'Oumuamua An Alien Probe?   150
    You Have a Visitor

Terraforming   154
    How to Engineer a Habitable Planet

CHAPTER 7
Do Aliens Do It Too?
WHAT WILL WE FIND WHEN WE FIND ALIENS?   159

Beyond Carbon-Based Life?   160
The Molecule of Love

Talking Tumbleweeds or Flying Forests   164
What Will Aliens Be Like?

Alien Minds   171
Can You Talk with an ET?

Alien Ethics   175
Should We Hide or Fire a Flare?

Will the Biological Era Be Short?   178
Welcoming the Robot Overlords

Ancient Aliens   182
How to Think about Million-Year-Old Civilizations

CHAPTER 8
Why Aliens Matter
IT'S MORE THAN YOU THINK   186

Acknowledgments   193

Notes   197

Recommended Reading   203

Index   204

About the Author   215

Introduction

Everybody loves aliens. I know this because everybody tells me they love aliens. Life in the universe is the first thing people ask me about when they hear I'm an astrophysicist. "Do aliens exist?" is one of those special questions, kind of like "What happens after you die?" Lots of opinions, no real answers, and, most important, actually knowing the answer would change the world.

The thing is: I love aliens too. In fact, I have been obsessed with them since I was a kid. I first got hooked when I found my dad's pulp science-fiction magazines as a five-year-old. On the cover of every issue were images of spaceships, barren moons, and bug-eyed alien monsters. From that moment on, I was on a mission to learn everything I could about the stars and alien life. This obsession made me a pretty annoying kid (apparently, I liked to quote the speed of light to four decimal places), but it also drove me to watch all the documentaries, bad sci-fi movies, and *Star Trek* reruns in existence. Any depiction of an alien was good enough for me as I dreamed of possibilities out there waiting to be discovered.

Back in the 1970s, at the height of my childhood obsession, the scientific search for life in the cosmos had barely begun. There were only a few very brave and determined pioneers carrying out the search for extraterrestrial intelligence (SETI), and most of them faced the scorn of their colleagues. SETI was considered a little "out there," marginal at best in the scientific community. A big part of that dismissal was just bias. There just weren't many astronomers who thought about the problem of life in its cosmic context back then. And it's true, we really didn't have much to go on in those days in terms of setting up a true scientific search for life among the stars, smart or otherwise.

Most of all, we didn't know if there were any planets in the galaxy other than the eight that orbited our Sun. This was a killer point, since scientists expect planets to be necessary to get even simple life started. So not having a single example of an extrasolar planet (an exoplanet) meant we literally didn't know where to look. We also didn't know much about how planets and life evolve together in ways that might keep a world habitable for billions of years, long enough for "higher" animals and even technological civilizations to appear. In short, when it came to searching for alien life in the universe, we were pretty much in the dark.

Not anymore.

As you read these words, the human species is poised at the edge of its greatest and most important journey. Over the past three decades, the scientific search for life in the universe—a field called *astrobiology*—has exploded. We've discovered planets *everywhere* in the galaxy, and we've figured out how and where to look for signs of alien life in the atmospheres of these new worlds. We've also looked deep into Earth's almost four-billion-year history as an inhabited world. From this view, we've gained new and powerful insights into how planets and life evolve together. Seeing the way life hijacked Earth's evolution over the eons gives us clues about what to look for on distant planets (like oxygen, which generally can exist in an atmosphere only if life puts it there). We've also sent robot emissaries to every planet in our solar system. With their wheels or landing pads on the ground, we've begun searching these neighbor worlds for evidence of life existing now or perhaps deep in their past. Most important, we have launched and are building insanely powerful, next-generation telescopes. With these tools, we'll finally

INTRODUCTION

go beyond just yelling our *opinions* about life in the universe at each other. Instead, we will get what matters most—a true scientific view of if, where, and when extraterrestrial life exists.

All these new discoveries, from exoplanets to Earth's deep history, are transforming what we think of as SETI. A new research field is rising that scientists are calling *technosignatures*,* which embraces the "classic" efforts of SETI while taking the search for intelligent life into new forms and directions. Knowing that the galaxy is awash in planets means we now know exactly where and how to look for alien civilizations. Rather than hoping for someone to set a beacon announcing their presence (one premise of the first generation of SETI), we can now look directly at the planets where those civilizations might be just going about their "civilization-ing." By searching for signatures of an alien society's day-to-day activities (a technosignature), we're building entirely new toolkits to find intelligent, civilization-building life. These toolkits will also allow us to find the kind of life that doesn't build civilizations. Using our telescopes to find a signature of a planet covered in alien microbes or alien forests (a biosignature) would also be a game changer in terms of how humanity sees its place in the cosmos.

So now, finally, we are on the road to finding those aliens I was so obsessed with as a kid. Or we're on the road to finding out we really are alone in the cosmos. Either answer would be stunning. It's a pretty damn exciting moment.

But it's also a confusing moment. Just as the scientific search for alien life is gaining steam, there's also been an explosion of interest in aliens that are supposedly visiting Earth right now. Over the last few years, a handful of videos taken by US fighter pilots have cropped up online showing fuzzy blobs appearing to fly in ways that would be impossible for normal aircraft. The videos have brought

---

* Some scientists still use SETI to refer to the field, and that's OK. But for many, including me, technosignatures correctly captures all that is changing in the study of "advanced" life in the universe.

xv

unidentified aerial phenomena (UAPs) into the spotlight, raising the stakes on the alien debate. But the UAP furor also confuses the issue about the giant leap science is taking as it begins looking for aliens in the most likely place (i.e., alien planets).

UAPs are the US government's new name for unidentified flying objects (UFOs), a subject that's been around for years, holding modern culture in thrall. UFOs as alien visitors make for great science fiction (everything from *The X-Files* to *Independence Day* to *Nope*). The possibility of their actual existence has mostly been dismissed by scientists. The overwhelming majority of astronomers see UFOs as natural phenomena that get misidentified, objects related to national defense, or just purposeful hoaxes. In 2021, however, the US government revealed more than a hundred UAP sightings for which it had no obvious explanation. The media tornado over the UAP videos was unrelenting, even as most scientists emphasized that *unexplained* can mean there simply isn't enough data, or good enough data, to even begin formulating an explanation. Still, in the wake of the new government interest, I am left wondering, "Do these things really have anything to do with aliens?"

Between the remarkable progress in astrobiology and technosignatures on the one hand, and the blizzard of coverage about UAPs on the other, aliens are big news. More than ever, we want to know: Is anyone out there? I wrote this book to help people understand that question as scientists see it, the definitive answers scientists are working to find, and, most amazing of all, how close we are to getting some of those answers.

For a chunk of my career as an astrophysicist, I studied less freaky-deaky stuff. At the University of Rochester, I ran a "computational astrophysics" research group in which my students and I used the world's most powerful computers to explore how stars form from giant clouds of interstellar gas and how they die by tearing themselves apart in titanic stellar winds. These were very cool projects, and I loved the vistas they opened for me. But I never lost my little-kid interest in cosmic life. So, about a decade ago, I started

INTRODUCTION

a research program in astrobiology, doing work on exoplanets and their atmospheres. Then I started thinking about climate change from the perspective of astrobiology, positing that maybe every civilization triggers its own version of global warming.

My life really changed, however, in 2019 when a group of colleagues and I were awarded NASA's first grant to study exoplanet technosignatures. That is, NASA began funding us to think about the best ways to look for alien civilizations. We applied for the grant because, over dinners (and beers) at international meetings, we all got way too excited (it was the beers) about those exoplanet discoveries and how they could rewire the search for intelligent life. But NASA had never funded a project like the one we were thinking of. In fact, after years of getting burned by Congress for funding SETI research as a waste of taxpayer dollars, the space agency had barely funded *any* work on intelligent life in the cosmos.

So when we put in our proposal, we kept our hopes low. But then to our surprise, amazement, and joy (and more beers), it was accepted. The frontier was opening. We'd been given a chance to help shape the most exciting quest humanity had ever taken on. It was a milestone for the field and a recognition of how much had changed in the scientific thinking about life in the universe. Since then, we and other researchers have been pushing into new terrain. We're all preparing for a truly systematic, scientific search for alien life and alien civilizations. That search is just getting started *now*.

It's from this vantage point that I see, and understand in my bones, why everyone wants to know about aliens. But if you're interested in the science—from SETI to astrobiology to technosignatures—where do you begin? There is a mess of history, concepts, and terminology floating around that you need to know to understand what's about to happen. What, for example, is the Drake equation, and why does it matter so much? What's the Fermi paradox, and how much SETI searching has actually been done to resolve the paradox? How many exoplanets are there, and which of them matter? What

A BRIEF GUIDE TO ALIENS

is a technosignature (or a biosignature), and how is anyone going to find one? And what about the UFOs/UAPs? Should we take them seriously? If we do, what are the questions we should ask, and how should we ask them?

The aim of this book is to give you a good ten-thousand-foot overview of what's happening now, what's going to happen soon, and why it matters so much. My biggest goal in writing it was to give you a fast, fun path into all the amazing questions and issues swirling around that mother of all questions:

Are we alone?

So, suit up. It's time to get started on our journey. We have a lot of ground to cover. By the time we're done, though, you'll have everything you need to know about everything there is to know (for now, at least) about aliens. From that point on, you'll be ready to join this great voyage of discovery, and you'll be ready when someone says we've found "them." Because in the end, we don't want to just believe; we have to know.

A BRIEF GUIDE

TO

ALIENS

CHAPTER 1

How Did We Get Here?

HOW OUR ANCIENT QUESTIONS ABOUT ALIENS TOOK THEIR MODERN FORM

Look at your hand. I know, it's a stupid request, but just look at it for a moment. Inside every cell in your hand and the rest of your body is the genetic memory of every ancestor, going back to the origin of *Homo sapiens* almost three hundred thousand years ago. That's more than fifteen thousand great, great, great, etc. grandparents. You carry multitudes within. You can also bet that all of those grandmas and grandpas stretching back through time spent some of their lives staring up into the clear night sky, with the sentinel stars staring back. And what does that mean? It means you're not the only one who's into aliens. Your parents were too. So were your grandparents, your great-grandparents, your great-great-grandparents, and so on.

OK, to be clear, maybe *your* parents or *your* distant ancestor in the fourteenth century wasn't obsessing about alien life. You can, however, be pretty damn sure someone else in each of those generations was thinking hard about it. That's because the argument about life in the universe is as old as arguing itself. "Are we alone?" turns out to be a really, really old question.

Debates about the existence of other inhabited planets go way, way back, and it's important to understand the form those arguments took, because they could get pretty heated. More important, those older debates lie as a kind of unspoken background for the great shift that happened in the middle of the twentieth century and the explosion of possibilities happening today. After the Second World War, the technologies of rockets, radio, radar, and atom bombs transformed how we thought about space and the possibilities of alien civilizations. It also brought about the first wave of widely publicized and recorded UFO sightings, which drove the idea of aliens deep into popular consciousness. In this first chapter we're unpacking that history so we can see exactly how we got to this crazy, amazing moment when a question as old as humanity is poised to get its answer.

A REALLY OLD QUESTION
Alien debates through history

We can track the debate about aliens, in writing at least, back to the ancient Greeks. Aristotle, one of the most famous of famous Greek philosophers, was what we'll call an *alien pessimist*. You may know of him, but in case not: he lived around 350 BCE* and developed sophisticated ideas about everything from the nature of art to the nature of biology. When it came to life on other planets, Aristotle was sure that Earth was entirely unique. That's because for him the Earth was literally the center of the universe: the Sun orbited Earth and so did the other five planets (the ones you could see without a telescope— Mercury, Venus, Mars, Jupiter, and Saturn). Aristotle saw the Earth as being so special that he divided the cosmos into a sublunar realm, meaning below the

---

* One good source for learning about the older history of these debates is *Plurality of Worlds: The Extraterrestrial Life Debate from Democritus to Kant*, by Steven J. Dick (Cambridge University Press, 1984).

Moon's orbit, and a celestial realm. Life and all its changes could only play out in the sublunar domain. The celestial realm was eternal and unchanging. It's from this perspective that Aristotle made his famous proclamation: "There cannot be more worlds than one." He meant that there can't be any place like Earth (with its unique life-forms) anywhere else in the whole universe.

Aristotle was, of course, an exemplar of big ideas at the time, and for nineteen centuries afterward. Eventually even the Catholic Church would adopt some of his views into its doctrine. But that doesn't mean all the other Greek philosophers of the Hellenistic period, about 2,350 to 2,050 years ago, agreed with him. There was, for example, a group of thinkers collectively known as atomists, who thought the split between sublunar and celestial realms was kind of stupid. For atomists like Epicurus, who lived around 300 BCE, everything everywhere in the universe was built from tiny bits of indestructible matter called atoms (atoms, atomists, duh). As these atoms sped around the cosmos, they collided and, in the process, combined in all kinds of ways. Here in our part of the cosmos, they collided to form the Earth and all of Earth's living stuff.

However, since atoms were everywhere and everything was made of them, Epicurus reasoned that the universe must contain lots of other planets, and many of them had to be inhabited. Nothing else made sense. If atoms were universal, how could Earth be special? That viewpoint made Epicurus an *alien optimist*, and he struck back at Aristotle's perspective, writing, "There are infinite worlds both like and unlike this world of ours. Furthermore, we must believe that in all worlds there are living creatures and plants and other things we see in this world."

Alien optimists and alien pessimists—while the details of these two positions are going to change over the centuries, it's pretty remarkable to see their basic outline appear in writing more than two millennia ago. That's about a hundred greats in the "your great-great-great-grandparent" way of describing the past. So, yes, people have been arguing about aliens for a long time.

• • •

After the fall of the Roman Empire around 500 CE, progress in astronomy shifted to the Islamic empires of Persia and elsewhere. Astronomers of the great Muslim societies continued to build on Greek astronomy, creating new and more accurate star charts, as well as adding new ideas about an Earth-centered universe of the kind Aristotle believed in. They, too, had their arguments between optimists and pessimists, adding Islamic theological perspectives to the split. Some scholars claimed the Quran supported the possibilities of other worlds and other humans. Then, as Europe emerged from the Dark Ages and began new scientific inquiries in the 1500s, the debate got even more heated.

In the late decades of that century, a radical Dominican monk, Giordano Bruno, appeared on the scene. Bruno was a Copernican, meaning he believed Polish astronomer Nicolaus Copernicus's view of the solar system with the Sun, not the Earth, at the center. Since all the planets orbited the Sun, that meant Earth had been demoted to "just another planet," nothing special. This *heliocentric* (Sun-centered) model stood against Aristotle's *geocentric* (Earth-centered) version, which also happened to be the official Church view. Copernicus knew the Church didn't take kindly to having its nose tweaked with astronomical heresies, so he held off publishing his theory until he was safely dead.

Giordano Bruno didn't think that way. He was both intellectually courageous and kind of an asshole, alienating almost everyone who supported him. Despite getting chased from one European capital to another, Bruno was willing to push hard on the limits of what the Catholic Church would tolerate in terms of heretical ideas. For Bruno, if you accepted that all the planets, including Earth, went around the Sun, you might as well accept that the stars were just other suns. From there he reasoned that all the stars had their own families of planets in orbit and some of these worlds must host life just as Earth does. The Church was very much of the "Oh no, they don't" opinion. Eventually the churchmen caught up with Bruno, and he was dragged before the dreaded Inquisition. For a variety of official reasons, Bruno was declared a heretic, hauled out into a square in Rome, tied upside down to a stake, and set on fire. Like I said, the alien debate has definitely gotten heated in the past.

Just eighty or so years later, the climate for arguing about alien life had cooled considerably. In 1687 Isaac Newton launched the modern version of astronomy with his drop of the great laws of motion and his description of gravity. By then, the heliocentric model had won out. All educated folks agreed that the Sun did not rise over the horizon each morning. Instead, it was the horizon that went down to reveal the Sun. In this new intellectual environment, a Frenchmen named Bernard de Fontenelle wrote *Conversations on the Plurality of Worlds* in 1686. The widely read book argued that the universe was teeming with planets and life. "The fix'd stars are so many Suns, every one of which gives Light to a World." De Fontenelle was most definitely an alien optimist. A century and a half later, the introduction of Charles Darwin's theory of evolution brought about an entirely new way of framing discussions of life and planets that favored alien optimism. Evolution had created life from rock and water and energy here on Earth, so why wouldn't it do the same elsewhere?

However, by the turn of the twentieth century, new discoveries in astronomy, not biology, made it seem that life might be astonishingly rare. That was because astronomers had come to the conclusion that planets like Earth might be astonishingly rare. By 1910 the dominant theory for planet formation required two stars to undergo a near-miss collision. The mutual gravitational attraction between the stars as they passed each other would drag material from both into space, and that material could then coalesce into a planet. But calculations showed that stars almost never ever collide. That meant planets almost never form. Since no planets implies no life, even as late as the 1940s, most scientists thought that aliens of any kind were impossible.

Big picture–wise, we've just covered 2,500 years of people asking the same question over and over again. During all that time, over all those generations, the question took on slightly different forms. Is Earth the center of the universe? Are there other planets? Did life form anywhere else? Throughout it all, though, what didn't change is the clue that nobody had. The alien question was just people on Earth debating each other, arguing to the point of setting each other on fire—over just like, their *opinions, man.*

In the fateful decade of the 1950s, however, things started to change. That's when, as we'll see, the first halting steps toward a *science of alien life* were taken. So, while questions about aliens may be really, really old, our capacities to answer the questions are really, really new. That's the story to be told in the rest of this book.

FERMI'S PARADOX
Is there a Great Silence?

The night sky is overwhelming if you get far enough away from artificial lighting. Deep in the mountains or out in the desert, the stars appear in all their glory, including the luminous band of the Milky Way stretching from one horizon to the other. But most of us never see this view, because we tend to live in the middle of the light-polluted cities we built ourselves. It's a weird state of affairs that so many humans in the modern world are denied this ancient experience of awe because of our need for artificial light. There's a vertigo, familiar to all our ancestors, that comes with looking into the blackness of space. There is a strange depth to a dark night sky. It's as if you can feel space's emptiness, stretching away to . . . somewhere. Most of all, though, comes a realization, uniting us with every human being who has ever stared into a crystal-clear night sky: "My God, it's full of stars."

The visceral sense of possibility inherent in the night and its seemingly infinite panorama of stars has driven much of our species' long

HOW DID WE GET HERE?

discussion about life beyond Earth. How could we be alone in the cosmos when I can see that there are so many other places life might form? But this argument is not much help in terms of getting a science of alien life started. How can you quantify "so many stars" in terms of *if*, *when*, and *where* life on another planet might originate? To do science, you need specific questions that can shape a specific research program. Without those questions, the arguments become too diffuse. It's hard to even know if you're making progress toward an answer.

As the last century reached its halfway mark, astronomy and physics had finally made enough progress to support scientifically relevant questions about intelligent alien life (i.e., technological civilizations*). While the very first such question was raised by accident, its origin story matters less than its impact. What we now call the Fermi paradox gave scientists one of their first well-formulated problems about interstellar alien civilizations. It, along with the Drake equation, which appeared ten years later and which we'll cover in the next section, brackets the remarkable decade of the 1950s.

Enrico Fermi was the kind of genius other geniuses spoke of in hushed tones. He could see instantly into the heart of scientific problems that left others flummoxed. This is exactly what happened with the question of alien civilizations. Late one morning in 1950,[1] Fermi and some colleagues were walking to lunch at the Los Alamos National Laboratory, the principal US nuclear weapons lab at the time. The conversation turned to UFO sightings, the first modern wave of which

---

* This is a good place to say we want to be mindful in our use of the word *civilizations*. Human beings have a long and tragic history of separating themselves into groups of "us" versus "them." Often the "us" are the ones who have civilization while the "them" are barbarians or savages or whatever epithet worked at the time. From my perspective, human beings have made many, many civilizations—from the amazing Seneca Confederation in what's now upstate New York to the Roman Empire to the Tang Dynasty in China. In this book, however, what we're interested in are the ways a technologically capable species organizes itself so that it will be visible across interstellar distances. That will involve the ability to harvest lots of energy, and that will probably require levels of technology we've only recently begun to achieve. That's what I mean when I write "civilization."

7

had begun just three years earlier. While the scientists were skeptical that UFOs had anything to do with aliens, they began asking themselves about the possibilities for extraterrestrial civilizations, including the technical details associated with interstellar propulsion—typical lunchtime physicist fare. After a while, the conversation drifted to other topics, and the colleagues just enjoyed their lunch. Sometime later, Fermi blurted out, "But where is everybody?"

In that instant over lunch, Fermi saw that if technologically advanced, star-faring civilizations really were common, they should *already* be everywhere, including Earth. It is a simple insight that takes our awe before the stars one step further. If life easily forms on other planets and easily evolves into advanced civilizations, Fermi asked, why hadn't aliens landed in Times Square and announced themselves? Why weren't they here already walking around and chatting us up on the street or, you know, taking over the government? Fermi knew that if aliens were common, they should be here now. How did Fermi know this? The answer is simple (once someone like Fermi shows it to you).

The logic, based on Fermi's insight, goes something like this: nothing can travel faster than the speed of light. That is a basic law of physics. It means that any alien spaceship will, in the best case, have to travel at velocities slightly slower. The Milky Way galaxy is made of billions of stars and stretches about a hundred thousand light-years end to end. Dividing distance (the size of the Milky Way) by speed (let's say about one tenth the speed of light, which aliens can probably muster if they're already flying around the galaxy) gives you the answer you need to Fermi's question. It would take hundreds of thousands of years for a technologically advanced civilization to hop across star systems (perhaps establishing settlements along the way) and cross the galaxy. Hundreds of thousands of years may seem like a long time to us humans with our short centuryish lifespans, but the galaxy is about 13.6 billion years old. That's *way* older. Relative to the age of the galaxy, any spacefaring species should be able to reach anywhere in the galaxy in the cosmic blink of an eye, yet we have not seen any definitive sign of them here on Earth. If technologically advanced exo-civilizations are common, then we should already have direct evidence of their existence.

That's where the paradox comes in: we have no such evidence. So, therefore... what? People have put a lot of thought into answering that question. Astronomer Michael H. Hart published a famous paper in 1975 on Fermi's insight, in which he argued that the absence of aliens in Times Square means that there are no aliens at all. For him, Fermi's paradox meant that we are the only intelligent life in the entire galaxy, end of story. With so many stars and no indisputable alien visitations, the only conclusion for Hart was that alien intelligence and maybe even alien life was incredibly rare.

Others haven't been so ready to throw in the towel. In fact, there is a whole cottage industry of scientific papers that try to explain away Fermi's paradox. For example, there's the zoo hypothesis, which says the galaxy might be full of aliens but they're purposely leaving Earth alone and watching us from afar, the way we view animals in a zoo. Another idea, championed by Carl Sagan, is that the expansion of spacefaring alien species always stalls quickly due to resource issues. The galaxy's entirety never gets settled and that's why there's no alien settlement here on Earth.

A few years ago, I was part of a research team that got in on the fun too. To get a better grasp of how spacefaring aliens could drive a wave of expansion across the galaxy, we created a way to simulate interstellar settlement. We began by assuming a single civilization evolved on a random planet orbiting a random star in the Milky Way galaxy. Then we modeled the civilization sending out ships crossing interstellar distances at a fraction of the speed of light. We assumed that once a ship arrived at a new star its crew would set up shop on one of the planets and build new ships. Those ships would then be sent out to other nearby stars. In this way, we were able to "watch" as a single civilization started an avalanche of settlements that swept across the galaxy. As Fermi predicted that day in 1950, our simulations showed that settling the entire galaxy would take just a few hundred thousand years to, at worst, a few hundred million years (still a short time compared to the age of the galaxy). Nothing, not even supernovas (exploding stars) that sterilize parts of the galaxy, could hold back the expansion. Even if a supernova did create a "dead zone" of uninhabited planets, settlement ships from outside the zone would quickly fill it in. The wave never stalled. Sagan was wrong.

Our models did show a way out of the Fermi paradox, however. As we know from Earth's history, civilizations don't last forever. The same would likely hold for alien civilizations, which could fail for reasons we understand like politics, war, disease, and natural disasters like supernovas. They could also end from causes we've *never* encountered, like subatomic super-parasites that turn digital technologies into zombie killers—I just made that up. Hold my drink while I call Hollywood and make the pitch.

The point is, if we allowed settlements to eventually die at a steady rate in our simulations, a new effect appeared. While there might be lots of aliens zooming between the stars at any given time, the fact that settled planets eventually went dark meant that holes were always opening in the map of the inhabited galaxy. If the conditions were right, in terms of stars being far enough away from each other, then those holes might not get filled in for a long time. That's how we found that relatively large pockets of space could be left sparsely populated for millions of years. If our solar system was in one of those empty pockets—a place that was empty during the relatively recent rise of our species—that would explain why there are no aliens here now. But it also means they could be on their way.

Of course, for those who believe that UFOs *are* alien visitors, there's no paradox at all, because "they" are here already. But this raises the question of why they're (mostly) hiding. The problem with UFOs as aliens, as we will see in the chapters that follow, is that you have to simultaneously explain why the aliens are always trying to hide and also why they suck so badly at it.

HOW DID WE GET HERE?

At the moment of his outburst over lunch, Enrico Fermi could not have known how important his insight would become. The Fermi paradox has haunted scientists, and anyone else who has tried to think about alien civilizations, in the decades since he made us consider it. While frustrating and maybe even depressing, the paradox has informed a lot of thinking about SETI by giving scientists a well-formulated question to work on. That means it's important for anyone interested in astrobiology, technosignatures, or UFOs to understand what Fermi's paradox is and what it says.

But it's also just as important to know what it does *not* say.

The Fermi paradox does not tell us that the universe is silent when it comes to alien life. It speaks only to the question of why the aliens aren't *here*, on Earth, right now. It doesn't say anything about our searches for signals from alien civilizations on distant planets.

Let me explain.

Sometimes people talk about SETI and the Great Silence. They say we've searched the entire sky with radio telescopes for over eighty years, and never once have we come across a message or signal or random TV transmission from extraterrestrials. But this is entirely wrong. The fact is, we've barely even begun to look. There is a misconception that every night, astronomers around the world probe the stars with their radio telescopes looking for alien signals. The reality is so much simpler and sadder. SETI has always been cash starved. Other than a few lone enthusiasts eking out a search here and there, when it comes to hunting for alien signals, not much has happened. The vast and overwhelming majority of the search that needs to be done *has yet to be done*.

There are a lot of parameters SETI searchers need to explore to complete a search. First, there are so many stars to look at. Next, they have to decide what radio frequency to dial into. Then there's the question of how often the searchers should look at each star: every day? every hour? every second? The list goes on, which makes it hard to keep track of who has looked at what. In 2018 Jason Wright and his colleagues at Penn State tried to do just that. Working on an idea from Jill Tarter, they examined every SETI search that had ever been carried out. They wanted to know what fraction of a full, systematic search had been completed. Their conclusion was pretty astonishing. If we imagine the amount of universe SETI researchers need to explore as Earth's oceans, Wright's team found that only one hot tub's worth of water had been searched so far. Would it make sense to declare the ocean fish-free after searching just one hot-tub's worth of it?

The moral of this story is that there is most definitely a *direct* Fermi paradox, but there is no *indirect* Fermi paradox. Yes, why haven't aliens set up a settlement here on Earth yet? You can come up with your favorite solution to that question because that question is the direct Fermi paradox. The solution can range from "there are no aliens" to "it's UFOs, bro." But when it comes to the *indirect* Fermi paradox, meaning evidence of alien life on distant planets, the answer is simple. There is no Fermi paradox of that kind. We just haven't really looked yet. Until now. For all the reasons we're going to cover in the chapters that follow, the search is heating up. We're ready. We're able. We're going to do it.

THE DRAKE EQUATION
Asking the right questions

Alien optimists look at a night sky filled with stars and are certain that what happened with life on Earth must have happened elsewhere. With so many stars, how could we be the only ones? Alien pessimists aren't impressed by the night because they're sure the odds *against* life forming overwhelm the number of stars. After all, if you bought one thousand lottery tickets but the odds of winning are one in ten billion, you're still going to lose (probably).

As we know, alien optimists and pessimists have been going after each other for most of human history, having nothing more than personal opinions to bolster their claims. But science doesn't run on opinion. It runs on research. If there was ever going to be real scientific progress on the question of alien life, somebody somewhere needed to give both sides something to actually work on. The Fermi paradox gave scientists a question, but it didn't give them a plan. Someone still needed to formulate a road map that scientists could follow in building an alien-life-hunting research program.

That person turned out to be Frank Drake, a soft-spoken radio astronomer from the Midwest whose courage was matched only by his creativity. His plan for a science of aliens became embodied in a simple equation formulated more than six decades ago. With it, Drake radically changed how we think about life in the cosmos and how we search for it. The Drake equation, as it is now called, remains the most important formula in astrobiology.

It was 1961, a little more than a year after Drake launched the first scientific search for alien life using a radio telescope at West Virginia's Green Bank Observatory. Drake was just over thirty at the time and worried that his interest in extraterrestrial life had earned him nothing more than ridicule from his colleagues. While that initial search hadn't gotten him the kind of accolades every scientist hopes for from their peers, it did earn him a remarkable call from the National Academy of Sciences (NAS). The guy on the other end of the phone was J. Peter Pearman, and he wanted Drake to lead a workshop on, of all things, "interstellar communications."

Pearman and the NAS wanted to explore the possibilities that we might soon be sending and receiving signals from extraterrestrial civilizations. Thankful that someone else was interested in cosmic life, Drake accepted the offer and began organizing the meeting. He and Pearman invited eight researchers to the meeting, including the renowned astronomer Otto Struve, the soon to be Nobel Prize–winning biochemist Melvin Calvin, the dolphin-language specialist John Lilly, and an ambitious young astrophysicist named Carl Sagan. As the opening day approached, Drake sat down to make an agenda to guide the discussion, but even this small task posed a unique problem. How do you organize a scientific discussion around a topic that had never been scientifically discussed? Addressing this problem is how Frank Drake created his now-famous equation.

His first step was to explicitly define the question. What exactly did scientists at the workshop want to answer? Naively, you might think it was "Do alien civilizations exist?" But that question didn't offer a clear way to drill down and figure out what steps—the all-important research plan—were needed to find an answer. And scientists do love a plan, so Drake seized on a different question instead. Rather than asking if aliens did or did not exist, he simply asked, How many of them are out there? It was a small change, and it changed everything.

Drake's exact scientific question was: How many advanced technological civilizations are there in the Milky Way galaxy? Drake defined *advanced* and *technological* as possessing radio telescopes that could be used for interstellar communications. Yeah, I know, that's a pretty narrow way to define civilizations, but it served a specific purpose: it led to a version of the alien question that could be broken down into a set of pieces. Each piece was a subproblem and led to a finer-tuned plan for scientists. By answering each subproblem step by step, researchers could build their way toward their coveted number of advanced technological civilizations, which we'll call N.

Discussing each subproblem separately also gave Drake the agenda for his workshop. Composing the problem mathematically, the most economical way for scientists to think, he combined all his subproblems into a single formula:

$$N = R_* \cdot f_p \cdot n_e \cdot f_l \cdot f_i \cdot f_c \cdot L$$

In words, this equation says (take a deep breath if you want to say it out loud; this is going to be a long sentence): The number of detectable alien civilizations in the galaxy equals the number of stars forming each year (R_*) times the fraction of those stars with planets (f_p) times the number of planets per solar system with a suitable environment for life (n_e) times the fraction of suitable planets where life actually does form (f_l) times the fraction of those planets where life evolves intelligence (f_i) times the fraction of those intelligences that go on to create technological civilizations (f_c) times the average lifespan of those civilizations (L).

You can inhale now.

Drake's little equation, scribbled on a blackboard more than sixty years ago, went on to become the foundation for thinking not just about *intelligent* alien life, but alien life of any kind. It has appeared in piles of scientific papers, been printed on T-shirts, been tossed about in movies and TV shows, and become the subject of countless podcasts and YouTube videos.

So what's the big deal? What does Drake's equation really tell us about alien life in the cosmos, and why has it endured? The answer is: evolution. What Drake's equation cleanly captures is how the appearance of technological civilizations is a cosmic process that starts from the simple and builds to the complex. To see this, let's walk through the equation's seven terms starting from the left and moving to the right.

The first term, R_*, focuses on stars, some of the first "structures" to evolve in the cosmos. The universe started out pretty simple right after the Big Bang, 13.8 billion years ago. There were just a few kinds of particles perfectly mixed together in a smooth soup of primordial gas. Stars were one of the first big "things" to evolve from that initial state. Beginning with the number of stars formed each year is the Drake equation's way of recognizing that stars are the universe's first step up the ladder from simplicity to the complexity of technological civilizations.

The second term in the equation focuses on planets. Planets are another step up in complexity. It is probably only on their surfaces or in their oceans that chemistry can work its magic and get life started. But when Drake wrote down his equation, no one knew if there were any planets out there other than the ones orbiting the Sun. It was entirely possible that nature hardly ever created planets.

That's why Drake's second term is the *probability* that any given star would have planets (written as f_p). If f_p was very close to zero, then planets would be rare, and that would be a bottleneck for the formation of life.

Not all planets are equal, however, when it comes to getting life started. Mercury is so close to the Sun that its surface is continually blasted by solar radiation. It's got daytime surface temperatures of 700° Fahrenheit. It's *very* hard to imagine how life could form in an oven like that. The opposite situation holds for frozen planets at the far edge of a solar system. Temperatures there would be so far below freezing that potential biochemistries would get starved of the energy needed to build something like a cell (i.e., anything complex). That's why the third term in Drake's equation is ne, the average number of planets per solar system in the band of orbits where the surface environment is neither too hot nor too cold. This is called the *habitable zone*.

The next three terms are where things start to get really interesting. Life is unlike any other physical system in the universe. Through the processes of Darwinian evolution, life is creative. It makes new forms in ways that stars, comets, and black holes never can. Where the hell did kangaroos come from? They're like weird giant rabbits that can punch you in the face. Comets never made a kangaroo and neither did any black hole, but, beginning with single-cell critters billions of years ago, life on Earth *invented* kangaroos through the process of evolution.

The first thing life must invent, however, is itself. In this way the formation of life represents a kind of big bang in the universe's capacity for complexity. The fourth term in Drake's equation deals squarely with the process called *abiogenesis*. That's a sophisticated name for the fact that life must somehow form from nonliving stuff (i.e., chemicals). Once again, Drake expresses things in terms of probabilities, with fl giving us the odds that abiogenesis will occur on a randomly chosen habitable-zone planet.

After abiogenesis, Drake focuses on the evolution of intelligence,

HOW DID WE GET HERE?

which has been slow going here on Earth. Life on Earth formed at least 3.5 billion years ago. The first animals didn't appear until almost 2 billion years later (about 500 million years ago), and the first possibilities for intelligence in animals didn't appear until brain size increased to something like what the dinosaurs had about 100 million years ago. If we're being conservative, we might say it wasn't until the bigger mammals appeared before intelligence could really get going. Drake expresses the question of intelligence as the probability fi that, once life emerges, it will go on to evolve into something like apes or dolphins or smart dinosaurs. Why is this important to consider? Well, you need some kind of intelligent species on your planet so that the next step, technological achievements like radio telescopes, can occur.

To build radio telescopes, intelligent species must develop largescale forms of cooperation. They must develop the means to create and store knowledge about the world. Then they must use that knowledge to extract and refine resources on large scales. In short, these intelligent life-forms must develop what we think of as a civilization. Of course, Rome, the Tang Dynasty of China, and the Aztecs were all highly complex civilizations, but they are not what Drake was talking about. Drake focused on radio telescopes because only civilizations that have reached *at least* our level of development matter for his question. (Remember, he was interested in interstellar communication.) Drake expresses this step in evolution as fc, the probability that, once an intelligent species appears, it will evolve into a complex technological civilization capable of building devices that can send messages across the insane distances separating stars.

The last term in Drake's equation focuses on something that goes beyond knowledge we already possess. We might not know how often planets evolve life that evolves into technological species, but we do know it has happened at least once. But how long, on average, does an advanced technological civilization like ours last? We have only been at the industrial civilization game for a couple of centuries, and we've had radio technology for only about one century. So do we have thousands, or even millions, of years ahead of us? Or will we flare out in just another century or so? Maybe other species would never be such jerks as to invent nuclear weapons. Maybe other species aren't so dense as to miss the fact that they'd triggered climate change and were wiping out most of their world's flora and fauna. On the other hand, maybe most species are even bigger assholes than we are. Maybe they are so mean, aggressive, and selfish that they take themselves off the galactic chessboard through general assholery in less than a century. These are the questions behind the Drake equation's final factor, the average duration of a civilization, i.e., the average lifetime (L) of civilizations throughout the galaxy.

By framing the question in the way Drake did, a workable path was forged to find N, the number of civilizations with which humans could communicate. It's worth noting that when Drake wrote down his equation more than sixty years ago, only a single term was known, the number of stars formed each year. We now know for certain that $f_p = 1$; that is, pretty much every star in the sky has planets. We also know that ne is about 0.2. One out of about five stars hosts a planet in the habitable zone where life can form. Coming up with the answers to fp and ne already is pretty amazing. It shows that, over just the last few decades, we've made spectacular progress on a question that's been around for more than 2,500 years.

How does the Drake equation work? In principle, we first get values for each term either by making measurements, constructing a theory, or just taking a wild guess. Then all the factors get multiplied together to get N, the total number of advanced civilizations in the Milky Way. Let's choose some numbers and see how this plays out.

We already know R_*, f_p, and n_e, so let's be optimistic about all the others and assume the other probabilities are 1.0 (i.e., every planet in a habitable zone forms life that becomes intelligent and creates a civilization). Let's also be optimistic about the average lifetime and assume L = one billion years. When we multiply everything out, we get N = two hundred million civilizations. That's a galaxy full of civilizations! There are four hundred billion stars in the Milky Way galaxy, so we'd have to look at about two thousand random stars before we found one with a civilization. That is easily within the possibilities of current technology.

On the other hand, a pessimistic estimation might put all the factors associated with life at, say, one in a thousand. It might also put the average lifetime of a technological civilization at just five hundred years. Since we've had radio for a hundred years, this gives us just about four hundred years until Armageddon, so smoke 'em if you got 'em. Put all these numbers into the Drake equation and you're left with N = 0—not even one civilization in the entire Milky Way. In this case, the galaxy is sterile. Of course, we exist, so we know this kind of pessimistic estimate is at least a little wrong.

Which case holds the truth? We can't say yet. But the Drake equation lets the optimists and the pessimists see exactly what they are being optimistic or pessimistic about. It lets them understand the explicit scientific questions (habitable-zone planets, abiogenesis, civilization lifetimes) underpinning the split between optimism and pessimism. Moreover, Frank Drake bequeathed to us a way to understand our own ignorance and organize a way out of it. It has motivated three generations of astrobiologists to get to where we are today, finally poised with the technology to answer his alien question once and for all.

THE UFOs ARRIVE
Kenneth Arnold sees saucers. Roswell gets busy.
The government reports.

Going from Fermi's sudden insight about interstellar travel times to Drake's equation providing the outline for an explicit research program, we can see a science of alien life and alien civilizations emerging in the early years of the Cold War. However, during that same period another version of the alien question made its appearance, and it, too, would leave a lasting mark on how people thought about life in the universe. It was just a few years before Enrico Fermi sat down to that famous lunch with his friends that UFOs muscled their way into the public imagination.

June 24, 1947,[2] was a good day for flying in the Pacific Northwest. The skies were clear and bright over Mineral, Washington. It was the middle of the day as amateur pilot Kenneth Arnold found himself navigating his small single-engine plane past the towering peak of Mt. Rainier toward an air show in Oregon. But he'd heard that a Marine Corps transport plane had gone missing and a reward was being offered for anyone who found its wreckage. Arnold decided to make a few circuits and have a look. He had no idea that he was flying straight into UFO history.

As Arnold surveyed the terrain below him, he saw a flash of light with a blue tinge. Was it sunlight glinting off another plane? He saw a DC-4 flying off in the distance, but there were no flashing lights coming from it. Then the flashes appeared again. This time he saw exactly where they were coming from: nine objects flying in a diagonal formation. Moving in unison like "the tail of a Chinese kite." Arnold watched as the objects banked and turned in ways that made him think he was watching some kind of advanced military aircraft. By using Mt. Rainier and Mt. Adams as a kind of yardstick, Arnold was able to estimate the speed of the strange aircraft. His little calculation yielded more than 1,500 miles per hour, almost twice the speed of sound. Because this was 1947, a few months before Chuck Yeager broke the sound barrier in an experimental rocket plane, the speed Arnold had just calculated seemed impossible. Arnold tracked the formation for a little while before the objects disappeared from view. The entire incident didn't last long but left Arnold with "an eerie feeling." After landing to refuel, he shared his story with friends at the airfield. What happened next would echo down history, shaping everything we think about UFOs and their connection to aliens from outer space.

Arnold's tale spread quickly, and reporters from the *East Oregonian* asked him to come in and give more details. To the newspapermen, Arnold seemed like a credible witness and a careful observer. Laying out the timeline of what he saw, Arnold described both the craft and their motions. Exactly what happened next remains controversial, but when Arnold described the objects as moving like "a saucer if you skip it across the water," he triggered a chain of events leading to what may be the greatest misquote in the history of journalism.

The story in the *East Oregonian*, a small paper by any account, ran with words "saucer-like aircraft." Then one of the reporters filed another story that was picked up by the Associated Press and somehow the description got garbled. Arnold's own account was of a flying craft shaped like a crescent with "wings" that swept back in an arc. That's not what other papers took from the AP wire story, which seemed to have misinterpreted Arnold's description. Then the *Chicago Sun*, not a small paper at all, ran the story with a spectacular front-page headline: "Supersonic Flying Saucers Sighted by Idaho Pilot."[3]

The *Sun* piece triggered an avalanche. Within six months, the flying-saucer story ran in over 140 newspapers throughout the United States. Even more remarkable, an epidemic of flying-saucer sightings began to sweep the nation. In the months following the story, more than 830 UFOs were seen in the United States and Canada, most of them being described, interestingly, as saucer shaped. By the end of summer 1947, "flying saucers" were officially a thing.

What did Kenneth Arnold really see that day? Beats me. I wasn't there. But you know what? Neither were you nor anyone else not named Kenneth Arnold. That's the first crucial lesson to be learned about his story. Arnold's tale was the original headline-making public UFO account. It set up an interplay between believers and debunkers that would come to define the UFO debate. Believers will point to the fact that Arnold was a respectable businessman who struck most people, including US officials who later interviewed him, as grounded and serious. That does seem to be true. No one at the time thought Arnold was trying to pull off a hoax, and his attempt to estimate the speed of the craft using the known distance between the mountains showed that he was a skilled pilot and thoughtful observer.

Skeptics, however, note that at least one official did say he seemed "excitable" (though why wouldn't he be if he really saw what he said he saw?), and then they offer reasonable non-alien things that Arnold might have actually seen. These include a mirage cast by a thermal inversion of the atmosphere, high-flying geese, or even a fracturing meteor. Believers then offer counter-arguments of their own, which, they claim, shatter these supposed explanations. Meteors, they might say, are never seen moving in the kite-like way Arnold described. And geese aren't known for flying at hypersonic speeds. This kind of back-and-forth then goes on until . . . well, until everyone else stops listening and moves on with their lives.

There is just not a lot that science can do with mere personal testimony, even from a witness who seems credible. In fact, as every cop and psychologist will tell you, personal testimony is the worst form of evidence. It has been shown time and time again to be wildly error-prone. Memory can inflate details that turn out to be untrue, or in their excitement, witnesses will miss things that would have been essential to an explanation. The best that one can say to someone like Arnold is "OK, I believe you are serious and that what you describe is what you believe you saw." If this seems weak, I'll point out that it's a lot better than "You're a lying dog who thinks he can take me for a sucker."

So, what *did* Arnold see? Were they spacecraft traveling at unprecedented speeds? Were they advanced military aircraft, still unrevealed after more than seventy-five years? Was it some natural phenomenon? There is no way to know and no way to agree that we know.

The second important lesson to be learned from Arnold is the power of a story. Arnold saw a flying saucer. He saw the first flying saucer. Then, almost immediately after the newspapers hit the stands, everyone else started seeing flying saucers. But here's the thing, Kenneth Arnold didn't see a "flying saucer." He saw a thin C-shaped craft whose back edge looked like those bat-shaped ninja star things Batman throws around . . . sort of. Then the newspapers misquoted Arnold, and that's why flying *saucers* are what people saw when they saw something in the sky. It simply is a matter of fact that a tidal wave of saucer sightings occurred *after* the Arnold story. Now, a believer might argue that's because no one took previous sightings seriously enough to publish them. But for skeptics, the rush of post-Arnold *saucer* sightings when Arnold didn't see a saucer makes it much harder to give weight to these new accounts, especially when they came in droves. That doesn't make those individual stories false; it just means there's a lot of chaff per grain of wheat.

Arnold's sighting began a critical thread in the tapestry of human thinking about aliens. It was the first real UFO (i.e., flying saucer) story, and was where the idea of technologically advanced interstellar life here on Earth *right now* entered the public consciousness. Regardless of whether you believe this is what's happening or not, the explosion of UFOs-as-aliens into global culture has had a profound (and mostly negative) effect on the scientific search for life beyond Earth. As we will see, the way UFOs appeared in our culture and the way they got handled, both outside and inside of government, didn't help in understanding what the hell UFOs really are. Basically, Arnold started an unintentional shit show, and nowhere did the poop land harder and pile up deeper than in Roswell, New Mexico. If there ever was a reason why serious scientists steer clear of UFOs and the culture that surrounds them, it is the head-spinning mix of true believers, hoaxes, and charlatans attached to the name Roswell.

Here are the basic facts of the case.* On June 14, 1947, W. W. "Mac" Brazel and his son were driving across the ranch they worked on when they found a field of debris composed of rubber strips, tinfoil, and sticks.[4] Brazel didn't have a phone or a radio at his place, so he'd missed the news from the week before about Kenneth Arnold and his flying saucer. But, after talking with folks in town, Brazel made a connection between the saucer hoopla and the debris. He spoke with the local sheriff, who in turn suggested that he contact the Roswell Army Air Field (now Walker Air Force Base). The Army assigned Major Jesse Marcel to look into the case. Brazel, the sheriff, and Major Marcel went back to the ranch and collected the debris. Marcel later made a statement to the local press that ran with the pretty stunning headline "RAAF Captures Flying Saucer on Ranch in Roswell." A few days after that, a statement from the War Department in Washington stepped the hoopla back, claiming that the stuff found on Brazel's ranch was from a weather balloon. The local paper ran a new headline: "Army Debunks Roswell Flying Disk as World Simmers with Excitement." Some pictures of Marcel and the debris were taken, and they most definitely *did* look like a pile of sticks and rubber, not an advanced alien spacecraft.[5]

And that was that. The story completely and entirely halted there, fading into history and obscurity. No one gave much thought to Roswell after that—except, of course, the Roswellians.

Thirty years later, the circus pulled into town.

Around the end of the 1970s, Roswell was dragged back into the limelight in interviews with Jesse Marcel by a UFO researcher named Stanton Friedman. From these came the claim that Brazel really *had* discovered pieces of a flying saucer on the ranch. The next step was an

---

* I am doing my best to give the "basic facts" here, but after reading multiple books with multiple storylines and multiple sets of characters, "basic" and "facts" seem pretty elusive. And that, my friends, is the heart of the problem with all things Roswell.

HOW DID WE GET HERE?

episode of the TV series *In Search of* . . . , dedicated to Roswell, the aliens, and an alleged government cover-up. Like a cancer, the story of Roswell was beginning to grow and evolve on its own, and with each step, the story changed as more witnesses and more details were added.

In 1980 Charles Berlitz and William L. Moore threw fuel on the fire when they published *The Roswell Incident*. These authors had previously scored minor hits with books about the Bermuda Triangle and the Philadelphia Experiment. The second book was about, you know, that time in World War II when a US Navy destroyer got caught in a time warp.

Roswell was the perfect project for these guys. According to their book, what really happened in 1947 was that a saucer flying near US nuclear weapons test activity got hit by lightning. This craft, which until that moment had managed to cross interstellar space unharmed, was somehow fatally wounded by the lightning strike, crashing on Brazel's ranch and killing all aliens aboard. *The Roswell Incident* included the first mention of alien bodies seen at the crash site, a detail that plays an increasingly important role later. Most important, the authors claimed they'd interviewed scores of witnesses.

The Roswell Incident was just the first in a line of books looking to mine the original story. Each new book added more alleged witnesses and more details, including the story of mortician Glenn Dennis getting a chance to view the dead aliens. Well maybe, or maybe not . . . Some of the books said there were more saucers and more aliens, some dead and some not. Two of these extraterrestrials, it was claimed, were even taken into custody by the government. There have even been claims that the alien bodies were viewed by none other than President Eisenhower.

It all gets very convoluted and, more important, very *not true* as has been demonstrated multiple times by multiple authors. The real problem is that no matter how many contradictions, absurdities, or hoaxes are found in these accounts, the faith of those who believe is not shaken. For scholars who study such things, Roswell is an example of the birth of a myth. It is a story that cannot be proven false. For the myth's believers, it exists outside the possibility of being proven false.

The story of Roswell must be mentioned for its effect on how UFOs are seen by the scientific community. We scientists have pretty high standards for what counts as evidence. The circus that played out with the Roswell story left anyone not part of the UFO true-believer community just shaking their heads. It also continues to cast a long cloud over discussions of UAPs today. Serious scientists who might want to take on the issue (without preconceived ideas about what UAPs are) must think long and hard before they enter the fun house–mirror world of UFOs. Why risk your reputation by getting caught up in that kind of tornado of the absurd?

But there's one piece of the story that really was a conspiracy. There was a government cover-up at work in Roswell. The weather-balloon story was indeed a lie. Instead, what crashed on Brazel's ranch was Project Mogul, a secret experimental program using high-altitude balloons to monitor Russian nuclear tests. We know this because in 1997, a full, detailed report was issued by *the government* called *The Roswell Report: Case Closed*, detailing the project and the cover-up at Roswell.

Real life, it turns out, has a sense of humor.

The tidal wave of flying-saucer sightings that followed Arnold's 1947 report did not go unnoticed by the government. Over the course of the next couple of decades, regular people and national governments entered into a UFO rinse-and-repeat cycle of fevered interest, high-profile reports, and finally, fading expectations. This cycle has clearly started again with the release of the now-famous US Navy pilot encounters with UAPs. The subsequent 2021 government report on sightings of lots of unidentified stuff in the sky raised the stakes higher instead of calming us down. NASA even convened its own UFO/UAP panel in 2022. To put these new efforts into proper perspective, we must go back and look at the long history of US government reports on UFOs. But get ready, the path here is not straightforward.

Just a year after the first flying-saucer craze in 1947, the US government launched its imaginatively titled Project Saucer to study the subject. Recognizing that that was a really bad name for an effort designed to ease public fears on the topic, it was rebranded as Project Sign. The official mission was to determine the nature of UFOs (i.e., if they were real) and if they posed a threat to national security. The emphasis on national security is something you have to pay close attention to here. It will help make sense of a lot that follows (if sense can be made).

Depending on your inclinations, it may not be surprising that sightings *and* government interest in UFOs occurred just as the Cold War was heating up. The United States saw itself in a deadly winner-takes-all conflict with Russia, a contest in which high-tech secret weapons and psychological warfare were the paranoid order of the paranoid day. Whenever the number of UFO sightings rose, along with media interest, the government became simultaneously concerned with two issues. First, were UFOs a threat? Did they represent technology that someone, terrestrial or alien, could use against the nation? The second question led in a different direction. Was the interest in UFOs *itself* a threat? Was the public's anxiety about UFOs something that could be used against the nation? The government's response to these twin concerns explains a lot about how we ended up with the shape-shifting, conspiracy-laced, perennial fog that is modern UFO culture.

Within Project Sign, a division emerged between those who thought UFOs weren't real and those who believed what came to be known as the "extraterrestrial hypothesis," or as author Sarah Scoles calls it, the "it's dumb vs it's aliens" dichotomy.* The disagreement had far-reaching consequences. As part of Project Sign, a document was supposedly written with the weighty title *The Estimate of the*

---

* I recommend Sarah Scoles's book, *They Are Already Here: UFO Culture and Why We See Saucers* (Pegasus Books, 2020) for a good overview of UFO history as well as what's happening now.

Situation, and according to Captain Edward Ruppelt (who led later UFO investigations), the conclusion of this thick top-secret report was that UFOs have an interplanetary origin.[6]

Whoa! There was a government report back in the early 1950s that acknowledged we were being visited by interplanetary aliens? Well, maybe. Or maybe not. Ruppelt revealed the existence of the report in a book written years later. But to this day no copy of *The Estimate of the Situation* exists in the public record. It's also worth noting that Ruppelt began the trend of former government and military officials writing tell-all popular books.

Eventually Project Sign was followed by Project Grudge (yes, that was a weird choice for a name). Its report concluded that most sightings could be explained, but somewhere around 23 percent resisted easy explanation. According to the report, "there are sufficient psychological [grounds] . . . to provide plausible explanations for the [sightings] not otherwise explainable."[7] In this way, the first official government report on UFOs concluded they (a) weren't real and (b) were not a threat to national security. What might be a threat to national security, however, was the potential hysteria associated with UFO sightings. Isn't it possible, the report seemed to conclude, that the Soviet Union could weaponize peoples' fear and fascination with UFOs as a form of mind control (a big topic back in those days)?

That governmental concern, not with UFOs but with UFO *culture*, seemed to become policy with the CIA's infamous Robertson Panel.[8] The secret four-day meeting convened by the spy agency in 1953 focused on the various ways that UFOs—real or fictitious— could be used by Russia to sow chaos or even mask an attack on the United States. The secret panel's secret report (not declassified until 1975) ominously implied that groups of private citizens who were coming together to investigate UFOs on their own should be watched to make sure they weren't being used for subversive purposes. Likewise, the report concluded that the government should make sure to publicly debunk *all* UFO reports and reassure the populace that the furor was all a big nothing-burger. The report even seems to discourage official reporting of UFO sightings, declaring that "in these perilous times [they may] result in a threat to the orderly functioning of the protective organs of the body politic."[9] So, yeah, the US government wasn't big on transparency about UFOs back in those days.

As the 1950s slid into the '60s, new waves of UFO sightings washed through culture and politics. In response, Washington initiated the next big government effort on the subject called Project Blue Book. Over the course of its existence from the 1950s to 1969, more than 12,000 cases were investigated. Of these, 11,300 were claimed to be explainable by the usual suspects.[10] Our sister planet, Venus, can be remarkably bright just around sunset and sunrise. Aircraft can appear stationary as they move toward the viewer until "suddenly" veering away. Balloons, birds, meteors, and known atmospheric phenomena all make their appearance in the list of explanations. In the end, only 700 sightings (about 6 percent) were said to resist explanation. And sometimes this resistance was simply about not having enough information to even begin forming a coherent attempt at an explanation.

In 1966, as Project Blue Book was chugging along, yet another uptick in UFO sightings led to the formation of a nonmilitary, academically based investigation (though its funding came from the government). University of Colorado–Boulder nuclear physicist Edward Condon was chosen to helm the effort. The idea was to give the Project Blue Book cases another, purely scientific look and address a simple question. Was there anything related to UFOs that was of genuine interest to science (like, you know, visitations by fantastically advanced alien civilizations)? After two years of work digging into the Blue Book files and attempting to parse out possible explanations, the Condon Committee released a report concluding that UFOs were, in fact, not of interest to science. The committee didn't see enough weird stuff to think some new phenomenon outside of well-understood physics was at work. Within a few months of its publication, the American Association for the Advancement of Science endorsed the committee's findings, and so the Condon Report became the de facto official scientific response to UFOs for the next few decades. If someone wanted to raise the issue of UFOs, they were referred to the Condon Report, and the subject was dropped.

Not everyone was on board with that assessment, though, and questions about the true scientific impartiality of the Condon Committee were quickly raised. In particular, the well-respected atmospheric physicist James E. McDonald was deeply critical of the Condon Report. McDonald was adamant that the committee's work (and the previous Air Force efforts) represented pretty weak science. Important evidence for the most interesting cases was not followed with any rigor, McDonald said, and key eyewitness testimony was never directly pursued. McDonald's criticism would echo down the years to today as calls for better, more rigorous science on UFOs got louder.

There are a couple of worthwhile takeaways from this story of government commissions, panels, and programs. The first is that only a fraction of unidentified things reported in the sky defy subsequent attempts at identification. The other is that during the formative first two decades of UFO sightings, the government really was less than forthright about its own concerns. It's entirely possible to be skeptical that UFOs have anything to do with aliens (which I clearly am) while also acknowledging that there was quite a bit of hanky-panky going on from the government's side. Whether the goal was keeping the Russians guessing about our technology or keeping them guessing about what we knew about their technology, Washington was A-OK with UFO disinformation.

However, by the 1990s, after the Cold War was over, new government reports come out that attempted to set the record straight about not setting the record straight back in the 1950s and '60s. In particular, a 1997 report admitted that the incident at Roswell was indeed a for-real government cover-up. In essence, it said, "Yes, we did lie about the 'weather balloon' back then, because it really was a secret atomic weapons detection program named Project Mogul."

The pro-alien UFO camp was not impressed. Clearly, US government efforts to allow the proliferation of UFO disinformation were purposeful but only temporary. They were a direct response to the geopolitical paranoia of the Cold War. So once that era wrapped up, at least some in government were going out of their way to say, "Oops."

There is a more general takeaway from all these reports, one that has held up over time. Most public sightings of UFOs have mundane explanations. But in the long history of UFO sightings, there have always been at least a handful of cases that, when looked into more deeply, do seem freaky enough to raise eyebrows. That doesn't mean they have anything to do with aliens, but it does mean they might be worth some more serious scientific investigation. The problem is that the mix of government obfuscation and UFO true-believer conspiracy mongering ended up bleeding over to SETI. As we will see, the incredulity associated with UFOs became guilt by association that almost killed hard-core scientific SETI efforts more than once. It's only now, finally, thanks to a series of stunning discoveries like exoplanets, that the whole field of astrobiology is climbing out of that shadow.

INVASION OF THE POP-CULTURE EXTRATERRESTRIALS
They're here!

If we want to understand the forces shaping astrobiology and the search for life *now*, we must look at more than scientific discoveries. The UFO sightings and the government reports weren't just news; they were forces changing society too. Scientists are human beings shaped by the cultures they grow up in. So are the politicians who fund science and the military commanders who worry about potential threats. This means there really are two kinds of aliens: those that may or may not live on distant planets, and the ones living in our heads.

Once the nuclear and space ages got going in the 1950s, aliens began their insidious invasion of our collective, cultural mind space. We didn't put up much of a fight. In the seventy years since Roswell, extraterrestrial life has been mainstreamed in everything from blockbuster Marvel movies to Ezra Klein podcasts. If we're going to think about where the scientific search for cosmic life is headed right now, we must also unpack how these other aliens ended up eating our brains. They say life imitates art and vice versa. That's doubly true when it comes to life in the universe. Everyone's ideas about extraterrestrials are colored in one way or another by the aliens that set up shop in our shared imagination.

Before World War II, extraterrestrials were hard to find. Outside of pulp science-fiction magazines like *Amazing Stories* or cheesy Flash Gordon serials, there wasn't much space for space or aliens in movies or radio (the dominant popular media of the time). The war, however, brought together two significant developments that made space and alien life part of popular culture. First, after the Soviet Union detonated its own atomic and thermonuclear bombs in 1949 and 1953, the instantaneous annihilation of all human civilization became a real possibility. Nuclear weapons demonstrated the godlike powers that technology could unleash.

Second, and perhaps more important for the aliens squatting in our minds, was the advent of powerful rockets. The Germans invented the V-1 and V-2 missiles that rained fear and death down on London. As the Cold War got hot during the 1950s, both the US and the USSR took rocketry even further, developing intercontinental ballistic missiles (ICBMs), which could toss warheads up to the edge of space and drop them back down on targets half a world away. ICBMs soon took on another role as the workhorses for the space programs of both nations. Earth's first artificial satellite, the USSR's Sputnik 1, was launched using a modified ICBM in 1957. Then, in 1961, the Soviets used the

HOW DID WE GET HERE?

same kind of rocket to blast the first human, Yuri Gagarin, into orbit. In 1962 the United States finally managed to score a significant space "first" by using a modified ICBM to drive the Mariner 2 spacecraft to Venus on the first mission to another planet.

On the one hand, people were freaking out about nuclear incineration while, on the other, they were witnessing the limitless frontier of space getting thrown open. It was a lot to deal with. To top it all off, breathless reports of UFOs were starting to pop up in newspapers. Hollywood, being Hollywood, read the tea leaves and decided aliens could be used to generate positive cash flow.

Beginning in the early 1950s, space travel and space-traveling aliens began to pop up as a mainstay of many studios' new offerings. Most of these were forgettable low-budget affairs with titles like *The Man from Planet X*, *Devil Girl from Mars*, and *The Beast with a Million Eyes*. You've probably never seen these movies, but I definitely have. But there were really thoughtful alien movies too, and they had a big impact. *The Day the Earth Stood Still* told the story of Klaatu, an ambassador from another planet. Landing a saucer-shaped (of course) spaceship in Washington, DC, Klaatu and his death-ray-wielding robot bodyguard had a hard message for humanity: mend your violent ways or stay out of space. *Invasion of the Body Snatchers* was the first alien invasion film. While the movie was also an allegory about political conformity (this was the era of witch hunts for supposed communist sympathizers), *Body Snatchers* was the first film version of the indelible and oft-repeated story of aliens as shape-shifting infiltrators. *Forbidden Planet*, the best of them all, gave us the idea of an advanced civilization falling prey to powerful forces, while also leaving its technology for later civilizations to find and use. This was a theme that was adopted by too many other science-fiction films, TV shows, and video games to count. Because I'm a nerd, though, I can't stop myself from naming a few: *Star Trek*, *Stargate*, *The Expanse*, and the awesome video game series *Mass Effect*.

Despite the abundance of cardboard spaceships and cheap alien costumes, the thoughtful, big-budget affairs ended up in the pop-culture mainstream. They had a staying power that has shaped our collective ideas about aliens and alien civilizations to this day.

Science fiction established a set of concepts and images about life in the universe that has had both positive and negative effects on our situation today. First, it made people aware that there were profound questions about extraterrestrial life that science might someday answer. The best of these efforts raised the questions in thought-provoking ways. On the other hand, all those terrible B-movie aliens made the whole subject seem like a joke. This was the birth of what's been called the "giggle factor" where any mention of SETI, technosignatures, or even astrobiology gets accompanied by raised eyebrows and snickers. As we will see, the giggle factor over "little green men" turned SETI into a political dynamite that almost killed its advance for thirty years.

So, when you think about aliens, the aliens you think about didn't appear out of nowhere. They are the fruit of many decades of collective cultural dreams in popular media. You need to keep those dreams in mind as we try to wake up to the new reality of a search for alien life that is poised to get very, very real.

CHAPTER 2

So How Do We Do This?

CRITICAL IDEAS THAT SHAPED, AND STILL SHAPE, OUR SEARCH FOR ALIENS

After all those centuries of just yelling at each other, people (well, scientists at least) finally got serious about finding aliens in the late 1950s and 1960s, when a set of foundational ideas about how to carry out a search for life beyond Earth was being established. Just as important, radically new ways of thinking about *how to think* about aliens were also being developed. Learning how to think systematically about how alien civilizations can evolve, behave, and be detected was a crucial step forward. It can be really hard to find something when you have no idea what you're looking for.

In this chapter, we're going to unpack some of these time-tested ideas. We're doing this for two reasons. First, if you're really interested in the scientific search for life beyond our planet, you just must know this stuff. Habitable zones, the Kardashev scale, Dyson spheres . . . they're going to come up in conversation again and again. Whether it's the science of exoplanets, the origin of life, or even discussions of UFOs, these concepts are the ground on which everything else is built, your foundation for smart-sounding things to say over dinner.

Of course, more important than impressing folks at the bar, the game, or the cocktail party is understanding why these ideas and this history matter so much. The answer is: science is conservative. I don't mean in terms of whom scientists vote for. What I mean is, science hangs on to its best ideas. It's very hard to figure anything out about the world, so when a scientist comes up with an idea that's useful or cleanly captures some complicated process like, say, the evolution of advanced civilizations, that idea has staying power. It will get used over and over again. It will get passed down to the next generation of scientists, who will also use it over and over again, integrating it into their thinking and their research practice. If an idea is really good, it may even become the cornerstone of plans for multibillion-dollar space telescope missions.

But it's not enough for an idea to be good. Eventually, evidence for that idea has to pass muster—that is, conform to what scientists call *standards of evidence*. Standards of evidence are the key. Everything depends on them. They're how the science sausage gets made. Unfortunately, standards of evidence are not very exciting. Nobody's inviting me on late-night TV shows to dazzle and entertain the audience with tales of proper procedures for obtaining accurate data. Still, more than black holes or quantum jumps or even alien life, standards of evidence are the most important idea in science, because they are the reason why science works.

Here's the thing. Creating modern science took a long time. We often hear the story of scientific progress through the work of singular geniuses like Galileo and Newton, but that's really just one part of it. For every Einstein, there were hundreds of people history barely noticed. Together they created networks of thinkers, mathematicians, and experimenters. They wrote letters that crisscrossed Europe and the world. They made visits to one another's workshops and laboratories. They created new institutions for

SO HOW DO WE DO THIS?

advancing science like England's Royal Society (founded in 1660). Most important, they argued furiously about the best way to carry out experiments on boiling liquids, new ways to derive formulas for tracking comets, and so on.

It was through these institutions and networks that the best practices for research were slowly and painfully worked out. Part of these best practices, perhaps the most important, were the rules for drawing conclusions from evidence. These rules tell you what counts as good evidence and what is crapola. They let you tell when a conclusion is supported by evidence and when it's a bong-hit fantasy that you really, really hoped was true but has nothing to do with reality. Why am I telling you about this? It's because standards of evidence are the difference between science and bunk. If we really want to answer the question of life in the universe, we are going to have to pay close attention to those standards so we don't get fooled. Recently, the astrobiological community has set down rigorous guidelines for what will count as a detection of life.[1] The SETI community pioneered such guidelines years before. Scientists spent their blood and sweat on these guidelines because we wanted to be explicit about what's needed before a research team can announce to the world, "We found it!"

Let's say my colleagues and I claim we've found evidence for city lights on a world fifty light-years away. The guidelines setting out the standards of evidence would demand that our signal be much stronger than the usual noise that occurs in any astronomical observation. They would demand that we look at all the ways our signal might have originated with our instruments and not the alien planet. Then the guidelines would demand that we look at all the other possible ways that kind of signal might have been naturally produced (perhaps by super-high levels of lightning on the alien planet). Only when all these boxes have been checked would we be able to make the claim that we'd found evidence for an alien civilization. Even then, we'd still expect our colleagues to come at us hard, challenging every aspect of what we did and what we claimed. That's just how science works, and on something as huge as the discovery of aliens, it has to work especially hard.

One of my biggest problems with UFO stuff is that there are no standards of evidence. So much of the time, any fuzzy picture or any witness claim is taken to be evidence of aliens among us. Of course, there *are* folks in the UFO community who do really care and do pay attention to standards of evidence, but too often they get drowned out. The noise is so much louder than the signal.

I start this chapter with a foray into the history and philosophy of science because, in the late 1950s and early 1960s, a handful of scientists *did* try to bring rigor to the alien question. Technologies had just emerged that made fledgling searches possible, and as scientists, they knew those standards of evidence were going to be the key to possible success. Nowhere is this story more explicit than in the heroic inaugural SETI effort pioneered by young, fearless Frank Drake.

PROJECT OZMA
The first search

There is a good reason why the search for life beyond Earth never got much attention in astronomy before the modern era. Simply put, it was impossible. Even the closest stars appear as nothing more than points of light in the best telescopes, and detectors sensitive enough to find planets around those stars didn't appear until the 1990s. The few astronomers who were interested in alien life back then had no way to even imagine how to scientifically search for it. Early on, there was a proposal to use massive fires in geometrical shapes to message intelligent creatures on planets within our solar system just to let them know we were here. Not surprisingly, setting Texas-size bonfires to flag down Martians never garnered much support. By the end of the 1950s, however, a true scientific search for alien life finally became possible, thanks to new technology and the plan drawn up by Frank Drake. It has taken over seventy years to get us to where we are today, poised to take Drake's work to the next level.

Doing astronomy with radio waves was still a pretty new idea back in the 1950s. Up until then, observing the sky meant using the kind of "optical" light your eyes were evolved for. But such light is just one small slice of the entire electromagnetic spectrum. All forms of light are electromagnetic waves traveling through space. The optical light your eyes use has a relatively short wavelength (the distance from one wave crest to another). Blue light, for example, has a wavelength measured in hundreds of thousandths of an inch. Radio waves are light too, but they have much longer wavelengths—the longest on the electromagnetic spectrum, ranging from the size of a fist to a high-rise, or longer. Beginning in the late 1940s, astronomers figured out how to build telescopes to observe the sky in radio waves. By the late 1950s the government joined the party, ponying up the funds to build the huge dishes that everyone now recognizes as radio telescopes.

In 1958, three years before he created the famous equation that bears his name, Frank Drake was still relatively fresh from getting his PhD in radio astronomy and was beginning a job at the newly commissioned Green Bank Observatory in West Virginia. Green Bank's radio telescope was still under construction when he arrived, and his colleagues planned to use their new giant device to study everything from the structure of our galaxy to exploding stars. Drake wanted to join these efforts too, but he was secretly hoping for another kind of target. When Drake was at graduate school, he'd met the famous astronomer Otto Struve. Back then, Struve was one of the few scientists interested in the question of other life in the universe. Unlike many of his colleagues, Struve believed that planets were common, and he thought some of those alien worlds might host their own forms of life. By example, Struve convinced Drake that searching for life was a valid problem astronomers could, and should, work on. Drake showed up at Green Bank wanting to take a crack at that problem himself.

The big radio telescope sitting outside Drake's window was the perfect tool for his ambitions. Because radio waves have such long wavelengths, they can travel through space without getting absorbed (blotted out) by interstellar dust particles the way optical light does. Not getting obscured by intervening matter means that even very faint radio signals can travel unimpeded across vast distances. Radio telescopes, even the early version Drake had, are also highly sensitive, meaning they can detect those very faint signals. So for Drake the next steps were clear.

He mapped out a plan by asking himself a straightforward question. How far away could Earth be seen by aliens using a telescope like the one Drake had access to? Using Earth as a baseline was good science. Rather than imagining some superpowered aliens, Drake began with what we knew existed, human levels of radio technology. When he did the calculation, he found that Earth's most powerful radio transmitters could be detected with a Green Bank–size radio telescope out to about ten light-years. That gave him a search radius. He next flipped the question on its head to get actual targets for the search. Are there any Sun-like stars within ten light-years? He framed the question this way to make the scientifically conservative assumption that life would require conditions similar to those on Earth. Because there was no way to know back then whether a star had Earth-like planets orbiting it (exoplanets would not be discovered until 1995), Drake did the best he could and focused on just Sun-like stars. He searched the charts for solar analogues within ten light-years and chose two. First was the star Tau Ceti in the constellation Cetus (the Whale). Second was Epsilon Eridani, a star in the constellation Eridanus (the River). Each was similar to the Sun in terms of size, mass, and temperature.

The next problem wasn't scientific but political. Could he get his colleagues at the observatory to buy into something as seemingly crazy as a search for alien civilizations? There was no field of

SO HOW DO WE DO THIS?

astrobiology back then. Astronomy was the study of inert (dead) celestial objects. Even contemplating the existence of life on other worlds was, for most astronomers, an unscientific flight of fancy. Why spend time thinking about a question you couldn't answer? The UFO craziness and the bug-eyed monster aliens of the popular B-movies of the time didn't help things, either. Most serious scientists wanted no part of questions about civilizations on other planets. The giggle factor was in full swing.

Luckily, the small group of astronomers at Green Bank were an open-minded bunch. Drake made his pitch for a search for extraterrestrial intelligence (i.e., signals of a non-natural origin) one afternoon over burgers and fries at a local diner. His colleagues liked the idea. Suddenly Drake had a real research project on his hands. He called the search Project Ozma after Princess Ozma in L. Frank Baum's novels about the land of Oz (Drake was a major Oz fanboy). Project Ozma played out back in the heroic days of scientific experimentation when often you had to design and build everything yourself. Drake also did it this way to avoid outside criticism that he was spending a lot of money on the project. But radio astronomy was a very gear-headed business back then. Like sound engineering at early blues and rock 'n' roll recording studios, the game was all about the electronics of tuners and amplifiers. But rather than capture that growl from Muddy Waters's electric guitar, Drake and his colleagues were trying to get a clean hold on faint emissions from a star ten light-years away.

Project Ozma lasted from April to July of 1960. It did not find any signals of alien origin. But the failure to find extraterrestrial intelligence did not mean that the project was a failure. As word leaked out, the entire world took notice. The young, lanky Frank Drake received a steady stream of visitors as he went about his ET-hunting business each day. Famous businessmen, leading theologians, and well-known journalists all came out to Green Bank to see what was happening. Drake and his project made it into headlines and TV shows around the world because, even then, the world cared about aliens. Suddenly this question of extraterrestrial life—discussed in bars, in schools, and around campfires throughout history and all over the world—had some hope of being answered. Project Ozma got the world's attention because it was represented on a threshold we were just beginning to cross.

Looking back, it's hard to overstate how significant Project Ozma was in terms of crossing that threshold or how ridiculously brave the young Drake was for taking us across it. There had never before been any search of any kind for any life beyond the solar system. Drake had to imagine the field into existence and give it a methodology that would adhere to those all-important standards of evidence. Once that threshold was crossed, there was no going back. Drake pushed all humanity from planetary childhood to cosmic adolescence. Suddenly we had the capacity, crude as it was, to find others like us in the vast ocean of stars. Suddenly we could do more than just dream about a conversation with other intelligences who, like us, gazed at the stars and asked how and why.

In the decades since Ozma, astrobiology has gone from fringe subject to reputable field of research. Throughout the 1960s and '70s, NASA sent missions to the planets in our solar system. With these efforts, there was some hope that maybe we'd find evidence of at least microbial life somewhere; Mars was the prime target. But by the 1980s, those hopes had mostly been dashed, and the small community that thought about even simple life in the universe remained small and not very well funded. The group of scientists who wanted to think about intelligent life and the possibilities of intelligence was smaller still. Although Ozma had gotten SETI going, the effort never became mainstream in the astronomical community. As we will see, part of the problem was that SETI became a political football in the 1980s. By the early 1990s, astrobiology was not near the frontiers of science, and SETI was very much at its margins.

Now, however, thanks to the revolutions we're going to be exploring in the rest of this book, the search for life has become one of NASA's highest priorities, with billions of dollars invested in building the next generation of space telescopes to make it possible. In 2020 the astronomical community under the auspices of the National Academies of Science produced their latest Decadal Survey on Astronomy and Astrophysics, setting the big priorities for new missions. What's sometimes called the Habitable Worlds Observatory—a mission to study exoplanets and find life—topped the list.* It will be the James Webb Space Telescope's successor in a couple of decades. That is the context in which Drake's achievement must be understood. It wasn't just the birth of SETI. Drake imagined and carried out the first true scientific observational astrobiological investigation. His tools were limited, and the astronomical understanding at the time imposed severe limitations on the kinds of search that could be carried out. But those limitations are gone now, blown away by the spectacular scientific advances of the last few decades. That's why we are finally ready to get the search going in earnest.

Drake's pioneering efforts with Ozma represent an epoch-making milestone. Many years from now, when perhaps we have a seat with all the other interstellar species at the Federation of Galactic Cultures, Drake and Project Ozma will still be remembered as the moment when it all began.

HABITABLE ZONES
Goldilocks in orbit

Frank Drake got the search for extraterrestrial intelligence started with project Ozma. However, science always works through an interplay between theory and experiment (or observation). It's not

---

* What this telescope will look like exactly and what it will be called are still up for grabs. There were competing proposals that the decadal committee looked at. What matters, though, is that the report went all in on building something focused on exoplanets and astrobiology.

enough to point a telescope at the sky and take data. Eventually, to become more sophisticated and raise the probability that your search is successful, you have to know what you're looking for. There's where theory comes in. It's the job of theorists to use physics to game out what is possible out there in the universe, which gets expressed in very beautiful mathematical laws. Seriously, they're as gorgeous as the opening bars of John Coltrane's *A Love Supreme* or Led Zeppelin's "Heartbreaker."

To build on Project Ozma, SETI needed a new body of theory. There had to be scientists providing theoretical concepts—ideas—that would shape searches beyond Ozma. In this section and the two that follow, we're going to unpack the key concepts that got laid down in the years just after Ozma was finished. Dyson spheres, the Kardashev scale, and habitable zones (this chapter) were the foundation on which future SETI searches would build and through which future SETI discussions would move. We could include the Fermi paradox and the Drake equation in this class of foundational alien search concepts too. These ideas continue to have a huge impact even now, as the classic Ozma-like radio SETI searches become just one part of the broader explosion of biosignature and technosignature studies. These key ideas also make an appearance when folks talk about UFOs as aliens. Those aliens had to come from somewhere, and these concepts are some of the best human thinking on that question.

We'll start with a simple observation about the frontiers of science. Sometimes making something up is the smartest thing a scientist can do. When Project Ozma was wrapping up in 1960, there weren't a lot of people trying to work out what a scientific search for extraterrestrial life should look like. In those days, if you were a scientist thinking, *We should do something about this question of alien life*, you were on your own. In response, the few researchers who cared did what scientists always do when standing at the edge of what's known. They made stuff up. They imagined a ladder and then climbed it.

SO HOW DO WE DO THIS?

The essence of good science is constrained imagination. Astronomers need guardrails to ensure that their thinking doesn't run off into the realms of the impossible or purely fantastical. To get other people on board with their ideas, researchers must take the laws of nature—like how gravity or chemistry works—and use them to build steps that can support their weight as they step higher into the unknown. This is the balance between having an idea and going out and getting data (observation/experimentation). Chinese American astrophysicist Su-Shu Huang took this path back in 1959[2] when he proposed that every star is surrounded by a *habitable zone* essential to the search for life. A habitable zone is the real estate within a solar system, described in terms of planetary orbits, where life can form. The idea tells astronomers where in an alien solar system they should spend their time looking for life.

What was the big constraining principle behind Huang's big idea? The answer is simple: water. Based on Earth's long history, life and water seem to be inseparable. When scientists look at how chemistry behaves, they see nothing else that works as well as water for life's needs. Water is the perfect solvent. So many chemicals dissolve in it that there's no equivalent to water as a medium for life's endless chemical shenanigans.

It is still absolutely *possible* that some life out there does use something other than water as its chemical foundation, and scientists absolutely need to explore this possibility. But for now, what matters is this: if you were starting from scratch, if you were building life from the ground up, water would be your best bet to get it going. That brings us back to Huang's golden idea.

The temperature on the surface of any planet depends on two factors: how far the planet is from its star and how much energy that star pumps out, which we'll call its luminosity. Let's take the distance part first. If a planet is really close to its mother star, that planet's surface will be scalding. It will be bathed in screaming torrents of stellar radiation day in and day out, making the planet a living hell (OK, a *nonliving* hell). If, on the other hand, the planet

45

is really far from its star, the planet's surface will be freezing cold and inhospitable, its star just a dim speck in the sky.

Now let's think about the stellar luminosity part of the problem. The luminosity of a star is set by how fast energy is being released by thermonuclear fusion in its core. Highly luminous stars pump out a huge amount of energy compared with a similar-size star with lower luminosity. That means a planet can be in an orbit pretty far from a very luminous star and still have a toasty warm surface. On the other hand, a planet orbiting a low-luminosity star will still be freezing unless it's on a crazy-close orbit. It takes Earth 365 days to orbit the Sun. Planets around low-luminosity stars will have habitable surfaces only if their orbits take an order of weeks to complete. An entire year in a few weeks or so . . . that's what I mean by a crazy-close orbit. If you want a comparison, Mercury, with its 700° Fahrenheit surface, is the closet planet to our Sun. It's got an orbit that takes 86 days to complete. To have comfortable temperatures on a planet around one of those cool stars would require that world to be about ten times closer than Mercury is to the Sun.

Putting this all together, if you give me a star and its planet and tell me the planet's orbital distance and the star's luminosity (its energy output), I can tell you the planet's surface temperature. I can tell you if the planet is a frigid snowball world, a searing desert world, or something in between, like our lovely planet. That's a pretty awesome piece of basic astrophysics.

Using this little bit of astrophysical magic, Su-Shu Huang figured out how to find the band of orbits around a star where liquid water could exist on the surface. The inner edge of the band is where the planet's surface is so hot that water will boil. The outer edge is where the planet's surface is so cold that water will freeze. On a planet anywhere in between, assuming it has an atmosphere, you can pour a glass of water onto the ground, and it will sit there in a puddle. In that little puddle maybe, just maybe, life could form.

The habitable zone was a simple, insightful idea, and like all simple, insightful ideas in science, it had the power to shape the future. Astrobiologists today use the concept as the basis for almost all their searches for alien life. Nothing makes an astronomer's heart beat faster than the discovery of an Earth-like world in the habitable zone. That's the big prize. That's what everyone is looking for. Habitable zones were one of the first guiding concepts in the search for life in the universe. It has stood the test of time and continues to tell us where those searches should be directed.

DYSON SPHERES
Aliens go big with megastructures

There aren't many terms in science that are more fun to say than "alien megastructure," and what's more remarkable is those two words have been showing up in the news recently in reporting on the hunt for alien civilizations. But what exactly is an alien megastructure, and why are sober-minded PhD scientists writing papers about them? The answer takes us all the way back to 1960 and a series of research articles written by the great theoretical physicist Freeman Dyson.

Dyson first earned his reputation by helping to develop the quantum theory of how electrons and light interact. This is the kind of thing that wins Nobel Prizes (which he did). But Dyson was also the kind of scientist whose curiosity took him far and wide, often outside the realm of ordinary scientific thinking. After Frank Drake began his search for aliens, Freeman Dyson began wondering what that kind of search might find. Being a theoretical physicist, Dyson tried to figure out where advancing technology might lead an alien civilization. Dyson's approach represented one of the first important attempts to use physics as a tool for imagining the trajectory of advanced technological civilizations.

Energy is one of the most basic concepts in theoretical physics. It's defined as the ability to do work. Anything that happens in the physical world requires the use of energy. Dyson saw that energy will always be the one limiting factor in a civilization's capacities. The more energy you have at your disposal, the more you can do. Because life probably requires the surface of a planet to get started, Dyson also recognized that every civilization will be born next to a titanic energy generator (i.e., a star). Ours is the Sun. A typical star produces about a hundred million billion billion watts of power. That's a million times more power every second than all the power plants on Earth produce in a year.[3] It would be natural, Dyson reasoned, for civilizations to eventually try to tap all the power streaming off their home star, and so he tried to work out the details of how they might do it.

As often happens in SETI, science fiction serves to inspire science. Beginning with ideas from Olaf Stapledon's 1930s novel *Star Maker*, Dyson imagined a structure made of orbiting light collectors that would surround a civilization's sun. These would basically be some form of solar panel that could harvest the star's power output. Capturing most of the Sun's energy would require a vast solar system–size array of these collectors or, in other words, a megastructure (though that term has come into fashion only recently). A single one of these megastructures would be a gigantic machine that could be far bigger than a planet. Dyson imagined a star surrounded by many of these machines catching stellar radiation and using it to power a civilization.

Dyson saw that the laws of physics, in particular the famous second law of thermodynamics, would cause the energy collectors to warm up as they processed the solar radiation. The second law says that any time energy gets harvested and used, there must be some waste. When you use a gallon of gas in your car, only some of that energy goes to do the useful work of moving the wheels. Much of it, however, ends up just heating the engine block. That heating, which doesn't do anything useful, is a consequence of the second law. In the case of Dyson's idea, the second law tells us that the collectors have to heat up as they catch and store starlight. As they warm, they also would start glowing brightly in the infrared part of the spectrum. Your body heat is making you glow in the infrared right now too, and you could see it if you had those military goggles that let you see in the dark.

Dyson's calculation showed that the infrared glow from such a starlight-collecting megastructure could be seen across interstellar or even intergalactic distances. In other words, if a civilization builds Dyson's energy-collecting megastructures, by the laws of physics, those megastructures will be detectable from their infrared radiation. We could see them. That's why this elegant idea caught on so quickly with SETI researchers—it gave them something to look for. Soon a new term, *Dyson spheres,* began popping up in other papers.

What do astronomers mean by a Dyson sphere? Think of a vast shell, like a Ping-Pong ball the size of Earth's orbit, surrounding and entirely enclosing the Sun. The inside of the sphere would be covered with solar panels or some other light-collecting technology. If the radius of the sphere is one Earth orbit, then its inside surface area would be around a billion times Earth's surface area. That's so much real estate that some researchers have suggested that civilizations might use a fraction of their Dyson sphere for living space. How many aliens could you fit on a billion Earths? It's a mind-boggling proposition.

Could anyone actually build a Dyson sphere? Is it even physically possible? Scientists love this kind of thing. There have been many papers over the years carrying out Dyson sphere calculations. They've found that to construct a shell one mile thick with a circumference the size of Earth's orbit, you would need to grind up all the mass held in all the solar system's planets. Then all that mass would have to be processed and used for fabricating the Dyson sphere's components. While this appears to be the alien equivalent of many trips to the hardware store, it is at least doable. A typical solar system *has* the material needed to get the job done. That conclusion meant that, while Dyson spheres might be the stuff of science fiction, they weren't the stuff of fantasy. Building one may require capacities we don't have, but it wouldn't violate the laws of physics.

Just one problem: Dyson spheres couldn't actually be spheres. After running through the physics researchers found that a giant hollow ball with a sun at the center would be gravitationally unstable. Give the Dyson sphere just a teeny nudge, and it slowly drifts to one side until it crashes into the star. While that would undoubtably be a bad thing, it would be very cool to watch.

Of course, any civilization with the technological chops to build one of these monsters would probably also know how to deal with the stability issues. Still, these problems have led most researchers (including Dyson) to conclude that a single Dyson sphere makes less sense than a Dyson *swarm*, a giant collection of independently orbiting machines that together wouldn't face the issues that come with building a complete enclosed shell. A Dyson swarm would still pump out infrared radiation, but it would not completely hide the star. That point will turn out to be important, because it is exactly why someone thought they might have found one around 2014. We'll get to that story later.

Dyson spheres were literally the mother of all alien megastructures. They were the first appearance in the scientific literature of an idea for large-scale artifacts of alien civilizations observable across interstellar distances. More important, the idea has endured. Even as the spheres morphed into swarms (or even Dyson rings), they gave astronomers their first cut at thinking about *astro-engineering*. Given enough time and enough technological progress, how far—how "mega"—might a civilization go? That's a question that still shapes the way we astronomers think today as our search enters its next and most promising phase.

• • •

THE KARDASHEV SCALE
How to measure an alien civilization

Dyson got astronomers thinking about engineering projects on solar system–size scales. That was an important step, because it opened the door to the next question: What does it mean for a civilization to become ever more "advanced," perhaps advanced enough to engineer a Dyson swarm or even go further? This is a question that any theory of alien civilizations must address. It's also a question that hovers over our own future—if we have one. How do technological civilizations evolve over thousands, hundreds of thousands, or even hundreds of millions of years? Tackling this issue is what gave SETI theorists another of their most enduring ideas: the Kardashev scale.

Given enough time and enough technological progress, what do civilizations like ours become? The industrial revolution on Earth is only a couple of centuries old, which makes thinking through this question a challenge for our imaginations. In the early 1960s, when SETI was just getting started, predicting alien technological evolution was recognized to go hand in hand with finding them. Once again, if you don't know what you're looking for, you're going to have a hard time finding it. It's a task my colleagues and I are still working on.

Russian SETI pioneer Nikolai Kardashev was the first to tackle the problem wholesale when, in 1964, he tried to formulate a big-picture view of how advanced civilizations evolve. While the United States had led the rush into SETI with Frank Drake's Project Ozma, scientists in the Soviet Union were not far behind. The Russians, having beaten the US into orbit, were just as space-happy and techno-optimistic about alien civilizations as their American counterparts. There was even some paranoia in the governments of both Cold War dance partners about who would make contact with aliens first and thereby gain some imagined military death-ray advantage. So just as in the US, Russian SETI had serious advocates doing seriously creative thinking.

Like Frank Drake, Kardashev was a radio astronomer. Being an observer by trade, he went into his project looking for a way to classify civilizations in terms of the signals they could produce (what we now call technosignatures). But like Drake and his equation, Kardashev and his classification scheme turned out to be far more important than his original intentions. The Kardashev Scale that emerged from his papers became an archetype for scientific thinking about the long-term evolution of civilizations.

Kardashev's scale was based on the belief that every technological civilization—whether built by savvy dinosaurs, smart octopuses, or sophisticated slime molds—passed through discrete evolutionary stages, which could be categorized. Kardashev could make this claim because his classification scheme wasn't based on social systems (were the dinosaurs authoritarian or democratic?), ethics (did the octopuses condone murder?), or political economy (were the slime molds capitalist or socialist?). There was no way to predict where aliens would fall on these dichotomies or if they'd even make sense to minds so different from ours.

Instead, like any good physicist, Kardashev built his classification scale on energy. It turned out to be an inspired idea. On a basic physical level, energy must underpin every civilization's evolution. No one, neither we nor aliens, can build a civilization and all its attendant infrastructure without using energy. Kardashev's idea was to identify the energy sources available to civilizations as they became more technologically capable. This allowed him to map the amount of energy available for use in each stage of development.

From Kardashev's perspective, there are three basic levels or types of advanced civilizations, based on their capacities to harvest energy:

Type I: *In the first stage of evolution, civilizations must harvest their planet's entire budget of energy. In most cases, the largest energy source*

SO HOW DO WE DO THIS?

available to a planet will be light from its star, so becoming a Type 1 civilization requires capturing all the sunlight falling on the planet. The Earth, for example, gets a few thousand atomic bombs' worth of energy from the Sun every second. Our minds tend to seize up trying to imagine so much power raining down on Earth all the time, but any Type 1 species would have all this energy at its disposal for civilization-building projects.

Type 2: *Why not go beyond just getting your planet's share of solar energy and capture the entire output of the home star? We've just seen how physicist Freeman Dyson famously anticipated Kardashev's thinking when he imagined an advanced civilization constructing a vast sphere around its home star. Each Dyson sphere would be a machine the size of the solar system built to capture all the light energy emitted by the star, rather than just the relatively tiny amount that falls on one planet. Type 2 civilizations can harvest billions of times more power than a Type 1 civilization.*

Type 3: *Next up is harvesting all the energy produced by all the stars in an entire galaxy. A typical galaxy has hundreds of billions of stars, so we are talking about a very advanced version of a very advanced civilization. This kind of Type 3 civilization might wrap Dyson spheres around every one of those stars, but once you get that advanced, it's possible that you are sucking energy out of space-time itself or some other science-fiction alternative.*

Where are we on the Kardashev scale? Do we have any hope of climbing even a little way to these Olympian heights? Carl Sagan and others have calculated where we sit on the scale, and it's not wildly inspiring, but it's not too shabby, either. By adding up all the energy we use from fossil fuels, renewables, and any other sources we have, Sagan and others have shown that we're currently somewhere around a Type 0.7 civilization. This number is, however, in logarithmic units, and while logarithms are supercool mathematics, they make me cry when I have to explain them. The important thing to know is the progression shoots up awfully fast, and that's the steep ladder that civilizations have to climb. We've still got a way to go before becoming a true Type 1 civilization. A recent calculation by researchers at the Jet Propulsion Laboratory, however, had us reaching Type 1 status in 2371 CE,[5] so invest in those solar-panel companies now.

The Kardashev scale is so useful because it gives us a ladder for our imagination. By formulating the question of civilizations and their evolution in a tractable way, we can get a handle on how all civilizations might progress and a baseline for what we might expect for our own future. Engineering entire galaxies, the realm of Type 3 civilizations, certainly seems like the work of gods. It's hard to wrap your mind around how a civilization with that kind of reach would organize itself, but Kardashev gave us a tool that at least lets us see that it's the kind of thing we need to consider.

Since Kardashev proposed this theory, others have picked up the ball, modifying or extending his scale. Other approaches have been brought to the table too. My research group has spent a lot of time thinking about the importance of *entropy* as well as energy. Entropy is connected to the second law of thermodynamics, which we met when exploring the Dyson sphere idea. As we've seen, when you use energy to do work, you generate waste of one form or another. Entropy is the name physicists give to that waste. Disorder is another name for it. Basically, anytime you use energy to do work and create order (like building a city, for example), you will also create some waste and disorder in the process. Build a city, and you create construction debris and a lot of noise. Both are examples of entropy.

On Earth we can see that climate change is the result of generating entropy by using so much energy for our own civilization building. The studies my research group completed showed that this might be a common problem. Any civilization like ours on any planet might

SO HOW DO WE DO THIS?

be forced through a climate bottleneck. Harvest enough energy, and the second law of thermodynamics creates feedback on your planet's climate system. Just like that, you get climate change. Our work used simple mathematical models of planets and civilizations to show that triggering climate change might be a universal step for advancing civilizations. Next, we can imagine that some civilizations are smart enough to make it through their climate crises. Others, however, may refuse to see, or act on the changes their energy uses are forcing the planet into. Those knuckleheaded species probably end up in the cosmic dustbin of history, a fate we will hopefully avoid. Either way, our results were directly applicable to Kardashev and his ideas.

Of course, Kardashev probably couldn't have known about our current climate crisis back in 1964. But rather than our results being at odds with the Kardashev scale, including climate change within it showed the scale's usefulness. By giving humanity a first cut at a physics-based, universal scale for the evolution of civilizations, Kardashev also gave us a general way to think about the overall problem. As time goes on and we learn more about how planets and civilizations interact (that's what climate change is really all about), we can add our new knowledge into Kardashev-like thinking. That's why the Kardashev scale has endured. That's why it remains such a powerful tool for imagining what had seemed unimaginable before: the advance of technological civilizations across cosmic time scales.

CHAPTER 3

WTF* UFOs and UAPs?

HOW THEY DO, OR DO NOT, FIT INTO THE SEARCH FOR ALIENS

Public fascination with UFOs began with the first major wave of sightings in 1947. The connection in popular culture between them and alien life was immediate. Because when people talked UFOs, what they really talked about were the aliens *in* UFOs.

This was not, however, a connection that the vast majority of scientists took seriously. Was this justified? Was there ever any reason for scientists seriously interested in the possibility of life beyond Earth to consider UFOs as a target for their studies? Even if the connection between UFOs and aliens might be weak, should UFOs have been of enough scientific interest to be studied simply because they were things that so many people saw that couldn't be identified? Whatever answers we might give to these questions in the context of history, we certainly can ask if those answers should be updated now, given the enormous amount of attention

---

* WTF stands for "What the Frack" because I'm a big fan of *Battlestar Galactica* (the reimagined series, not the original from 1979, which was embarrassingly stupid but did have great post–*Star Wars* special effects).

that UAPs (unidentified aerial phenomena,* the official name the US government has given to UFOs) have garnered over the last few years.

Personally, I picked up a highly skeptical take on UFO sightings based on reading works by my superhero Carl Sagan when I was an alien-obsessed but also *science*-obsessed kid. Sagan's oft-repeated mantra that "extraordinary claims require extraordinary evidence" made a lot of sense to me. I had read about those government reports we explored in the first chapter and didn't see any reason to question what the Condon Committee had found. More important, as a kid I'd also read Erich von Danikien's *Chariots of the Gods*, which made a huge impression on me. For a while I was totally sure ancient aliens had visited Earth and went around telling everyone as much (my poor dad endured a lot of these conversations). Then I saw a PBS documentary called *The Case of the Ancient Astronauts*, which had actual archeologists looking at the actual data concerning the building of the pyramids, the stone heads on Easter Island, and those animal-shaped lines on the Nazca plains. In every case, von Danikien's claims turned out to be inane. I'd been taken for a rube and that, for someone from New Jersey, is the greatest sin. The debacle definitely left me feeling burned by folks making extraordinary claims with crappy data or, worse, complete horsepoop.

Sagan's demand rang especially true as I got older and started learning the process of science in earnest in college and then as a graduate student. I was impressed (staggered, really) when I learned that the laws of physics allowed me to take the temperature of distant stars just by noting their color (red stars are cool, blue stars are hot). Along with learning the profound power of physics for revealing the structure of the world, I was also being introduced to those all-important standards of evidence. From where I stood, the

---

* Sometimes we also see "UAP" as an acronym for "unidentified anamolous phenomena," but jeesh, how many times can they change the name? We'll stick with the "aerial" for now.

swirl of fuzzy photos, hard-to-believe stories, and the penchant for conspiracy theorizing associated with UFOs meant that it was not a sandbox I wanted to play in.

But what about now? With the government admitting that its own Navy pilots are routinely sighting stuff in the air that can't be identified, is it time to update scientific attitudes? And if those attitudes are updated, then what, exactly, should happen next?

In this chapter, we're going to poke around in the history a bit more to understand why UFOs got the scientific bad rap they did, reexamine those older government reports, and then dive into what's happening right now with the UAP sightings and their consequences. Our first task, though, is to understand what effect UFOs had on SETI. The short answer is, they almost killed it.

THE GIGGLE FACTOR
How politics and UFOs almost killed the search for alien life

UFOs helped make SETI an easy target. Years of dubious claims, full-on hoaxes, and the Roswell-esque penchant for tin-hat conspiracy mongering created a cultural background that meant any discussion of extraterrestrial biology in serious company was guaranteed to raise eyebrows and trigger snickers. Alien life became a fast joke for late-night TV shows.

During the early SETI years in the 1960s and '70s, pioneers of the field like Frank Drake, Carl Sagan, and Jill Tarter soldiered on through these jokes, fearlessly establishing the scientific basis for a search for life out there. For a while, the government had their backs, but as time went on, that support began to crumble. A handful of politicians saw that they could use the stink of UFOs and B-movie bug-eyed aliens to raise their own visibility by attacking SETI.

In those heady first decades, however, a number of government

WTF UFOS AND UAPS?

science agencies had a healthy interest in the search for extraterrestrial life. NASA wanted to go microbe hunting on the other planets in our solar system, if they could be reached. And regarding intelligent life, it was the National Academy of Sciences that hosted the Interstellar Communications meeting where the Drake equation was born. As the 1960s turned into the '70s, SETI scientists also worked with NASA in ways that went beyond radio astronomy searches for alien beacons. There was consideration of Project Cyclops, a giant array of hundreds of radio telescopes sensitive enough to find unprecedentedly faint signals of intelligent life among the stars. And SETI scientists were helping NASA plan new telescope technologies for hunting exoplanets.

Then the politics showed up.

William Proxmire was a Democratic senator from Wisconsin who liked to think of himself as a fiscal hawk. He took it upon himself to bestow his so-called Golden Fleece Award on agencies that funded anything he considered a waste of US tax dollars. Because most of the science projects he targeted got only meager amounts of money, Proxmire's award was basically clever politics aimed at targets who couldn't fight back. In 1978 NASA's small portfolio of SETI funding fell into Proxmire's crosshairs. He gave SETI the Golden Fleece Award and, being a powerful and influential senator, he got his colleagues to keep the agency from providing any new funding. Proxmire relented only after Carl Sagan, by then a famous and well-respected scientist, publicly intervened, meeting personally with the senator to discuss the issue.[1] While the ban on SETI funding was eventually lifted in 1983 the public political flogging of SETI as wasteful kookiness had begun.

Even though NASA's SETI funding remained minuscule in the post-Proxmire period, it was still a target. In 1990 NASA tried to ramp up its SETI funding, from $4 million to $12 million, for a new search in the microwave region of the electromagnetic spectrum. While this is less than chump change in the federal budget, some politicians once again smelled blood. Making the link to UFOs

59

explicit, Republican Congressman Silvio Conte of Massachusetts tried to kill the funding, claiming, "... we don't need to spend $6 million this year to find evidence of these rascally creatures. We only need 75 cents to buy a tabloid at the local supermarket."

The same game was played again a few years later. In 1993, $12 million were finally allocated for the new search. Not wanting to attract more congressional attention, the project was stealthily called the High-Resolution Microwave Survey. Unfortunately Richard Bryan, a Democratic senator from Nevada, still caught wind of the effort and saw it as an easy chance to make some headlines. He sponsored an amendment killing the project, announcing that it would be "the end of Martian hunting season at the taxpayer's expense." Of course, Bryan knew that NASA wasn't planning to turn its telescopes toward Mars, but who cared? His quip made for great copy and linked SETI to the cultural fringes where all UFO enthusiasm lived, from the serious to the silly. The giggle factor had killed the search for life elsewhere in the universe again.

In the wake of these very public whippings, NASA learned the lesson that SETI was political poison. Although SETI scientists like Frank Drake and Jill Tarter did their best to show that the field lived within the necessary scientific standards of evidence, the damage was done. While the agency did what it could in the decades that followed, it became an accepted truth among researchers that federal support was going to be hard to come by. Still, SETI scientists soldiered on, raising private money where they could, but for all their best intentions, the search was running on fumes. The giggle factor had won.

Choking off SETI funding had important consequences for the search for life in the universe, because, basically, it meant there was no search. Using big telescopes costs money . . . big money. If there was no funding for SETI, then no telescope time would be granted for SETI. The political temperament about it that's held sway for most of the last seventy years means that our sky has effectively remained unexplored.

It's impossible to deny the role that UFOs had in the development of this history. As historian Steven Garber put in an article about SETI and NASA, the field "always suffered from a 'giggle factor' that wrongly associated it with searchers for 'little green men' and 'unidentified flying objects.'" Because of this association, astronomers never got the chance to get a real search started. Now, in the wake of exoplanet discoveries and powerful new telescopes, everything has changed. Radio SETI has become one part of the new field of technosignatures. Just as important, NASA has jumped into the search for biosignatures in a very big way. At least one question remains, though: Has anything changed for UFOs now that they, too, have gotten a new name? To answer that question, we still need a better view of history, so we can see the wheat, the chaff, and the three-card monte.

HOAXES AND HOAXERS
A good con never dies

People are pulled into the subject of UFOs for many reasons. Some just want to know what's behind the truly unexplained sightings. They're honestly looking for answers, be they alien or domestic. For others, the aliens who supposedly fly UFOs take on an almost religious aspect, providing answers to our big, unanswerable questions. But for some, however, the swirling mix of credulity and the incredible represent fertile ground for a con. Whether for money, fame, or a good laugh, the history of UFOs includes a fair number of famous hoaxes. The fact that true believers remain adamant even after the hoax is revealed is an important thread in the story of why scientists have not taken the subject seriously. Life is short and good scientific research takes many years to many decades. That means researchers must literally make bets with their lives (their working years) about what is worth their time.

One of the hallmarks of those all-important standards of evidence is that good data must be able to kill bad ideas. As a scientist, I train my PhD students to let their favorite theories die if observations demand it. If mounting evidence stands against your cherished idea, let the idea go and move on. In this way, science basically teaches us when and how to change our minds. That's a really beautiful thing. It's also not what happens with these UFO hoaxes.

UFO cons began to appear almost as soon as UFOs. One particularly noteworthy case involved the town of Aztec, New Mexico (why New Mexico figures so prominently in UFO stories is anyone's guess). In the late 1940s and early '50s, as the UFO craze was just getting started, Silas M. Newton and Leo A. GeBauer were traveling the country claiming to have artifacts recovered from a saucer that crashed outside Aztec. The two men were looking to sell the artifacts, which, they said, were made of a metal retrieved at the crash. They claimed machines they'd fabricated from their alien metal could find underground deposits of oil, petroleum, or even gold. It was through Newton and GeBauer that *Variety* columnist Frank Scully got hold of the story. Scully first published an account of the Aztec UFO crash in the magazine, and later in his bestselling book titled *Behind the Flying Saucers*. As he told it, the crash occurred in 1948 twelve miles from town. Not only was the saucer recovered, wrote Scully, but alien bodies and even their concentrated food wafers were too. The military was covering the whole thing up, of course, including the craft's Venusian origin and the "magnetic principles" it worked on.

It was all pretty ridiculous stuff. Venus, for example, is an uninhabitable hellhole of a planet with daytime temperatures of 700° Fahrenheit and surface atmospheric pressures that could crush a nuclear submarine. Not a good place for life. What matters though was the cachet the *Variety* story gave Silas Newton and Leo GeBauer, both named as sources in Scully's writing. The two used this minor fame to sell their fake alien artifacts to a variety of rubes. Eventually, the hoax was revealed, the victims spoke out, and both men were convicted of fraud.

But, as with many things UFO, even the convictions could not kill the story. There remain folks convinced that a saucer did crash near Aztec. Their confidence rests mainly on a memo buried in FBI files that simply recounts an unconfirmed report of saucers recovered in New Mexico. The agency never regarded it as important enough to follow up on, but the so-called Guy Hottel memo, named after the agent dutifully passing the report on, remains a hot-button issue for some UFO true believers.

A more modern hoax also revolves around New Mexico, but this time the nonexistent crashed saucers were the ones that didn't crash outside of Roswell. In 1995 the Fox TV network aired the documentary *Alien Autopsy*. The show focused on a grainy film that purported to be actual footage of actual government scientists hovering over two actual alien bodies. These were, it was claimed, the extraterrestrials recovered from the Roswell crash. A British entrepreneur named Ray Santilli said he'd recovered the film from an aging cameraman who'd been present at the autopsy. The footage was pretty graphic, which drove its appeal. The aliens, with their triangular heads, gaping mouths, and big dead eyes, are naked (yikes) and show a fair amount of damage, ostensibly from the crash. It was the stuff of nightmares, really. The program was so successful, Fox ran it three times.

Alas, all was not well at the alien morgue. Several suspicious reporters tried probing details of the story, but it wasn't until more than ten years later that another documentary managed to fully crack the case. Over the course of an uncomfortable hour of TV, Ray Santilli admitted that his film was a "reconstruction." The alien bodies in the film had been sculpted using casts filled with sheep brains, raspberry jam, and chicken entrails (yum!). And the secret 1947 government lab was really somebody's living room in 1995 London.[2] After the scenes were shot, the fake alien bodies were cut up into pieces, à la *The Sopranos*, and dropped into dumpsters around London.

But even after revelations like these, the stories don't die. You can still find people who will adamantly tell you that a flying saucer crashed in Aztec, New Mexico, and that *Alien Autopsy* represented a real event. Of course, people are free to believe whatever they want for whatever reasons (as long as they aren't hurting anyone). But this kind of credulity is what's made UFOs toxic for people in the scientific community. Any scientist who might want to investigate a truly compelling case has to explain to colleagues, "Yes, all that other stuff is donkeycrap, but here, here is a case that really does merit attention."

Good luck with that.

Even if you can get past such a hurdle, there's also the reality of getting bombarded by true believers who demand that you acknowledge that *Alien Autopsy* is real and aliens exist on Earth. This is the foundation, set to dry over decades, that you must know before you can understand the modern era we live in now, in which UFOs have turned into UAPs and the government is all in on their existence.

THE McDONALD CRITIQUE
So, about those unexplained cases . . .

In spite of the hoaxes, the conspiracy theories, and the towering pile of fuzzy-blob saucer pictures that science can't do anything with, there are UFO sightings that remain truly remarkable and remarkably weird. Since we're about to launch into the story of the modern frenzy in which UFOs became UAPs, it's worth remembering some of these cases and a scientist named James McDonald who adamantly refused to let the scientific community forget them.

It started as just another routine training mission for the crew of the RB-47 bomber. The six Air Force men aboard were all a few days away from being reposted to Germany from Forbes Air Force Base in Topeka, Kansas. The date was September 17, 1957,[3] and the bomber was flying high through the darkening skies of Mississippi. As the plane made its way along the coast, the radar operator picked up the trace of an object miles out in the gulf. A little while later, the pilot and copilot saw what appeared to be the landing lights of a small craft flying ahead of them. Then, without warning, the lights winked out. The puzzled crew radioed ground radar stations to see if they were picking anything up. The immediate answer was yes, ground radar said there was *something* flying in the sky with them.

Then, suddenly, a blinding red light "about the size of a house" appeared in front of the bomber. It hung there, traveling with the plane for almost a minute. Just as suddenly, the light blinked out. The radar traces disappeared as well. The stunned crew began to circle the plane, trying to find whatever "it" was again. Unexpectedly, the strange light made yet another appearance. It snapped back on, following a course aligned with the bomber's original flight path. Radar traces also reappeared on both the plane and the ground systems. After another ten minutes, the light and the radar traces disappeared completely, this time for good.

Taken on its own, this story is definitely goose-bump worthy. More important, it figures prominently in a critique of the famous Condon Report we covered back in chapter 1 by a well-known and, at the time, well-respected physicist, James McDonald. Before entering the UFO fray, McDonald had established the Institute for Atmospheric Physics at the University of Arizona. He was a guy whom people in the atmospheric physics community knew and had confidence in. But McDonald's interest in a broad range of topics eventually led him to investigations of UFOs as atmospheric phenomena. He studied UFOs for years, sometimes embracing and sometimes rejecting the alien hypothesis. McDonald was consistent, though, in believing that UFOs warranted deeper scientific study. From that perspective, he became a vocal critique of the Condon Report, which had become the definitive scientific study of UFOs at the time. In 1969 McDonald spoke at a conference on UFOs hosted by the American Association for the Advancement of Science, where he gave a blistering rebuke of the Condon Committee's treatment of the sightings listed in the "unexplained" category. In McDonald's own words:

> No scientifically adequate investigation of the UFO problem has been carried out during the entire 22 years that have now passed since the first extensive wave of sightings of unidentified aerial objects in the summer of 1947. Despite continued public interest, and despite frequent expressions of public concern, only quite superficial examinations of the steadily growing body of unexplained UFO reports from credible witnesses have been conducted in this country or abroad.

McDonald argued that both the Air Force and the Condon Committee did a shoddy scientific job in their analysis of the most puzzling UFO cases. McDonald was initially supportive of the Condon Committee's work, and he had access to all of Project Blue Book's files, heightening the credibility of such claims. In addition, he took on the task of going back to find and interview many of the witnesses involved in these cases.

McDonald turned his speech into a paper called "Science in Default."[4] I was given a copy by two members of our NASA technosignature team, Jacob Haqq-Misra and Ravi Kopparapu. Both are excellent scientists, and both believe that a purely agnostic approach to the study of UAPs is warranted. After reading McDonald's article and looking into his career, my own position softened. My experience with things like the conspiracy-laden "anything equals evidence" sinkhole that is Roswell had led me, like many scientists, to steer clear of anything related to UFOs. But I found "Science in Default" to be a strong argument that the previous government reports were lacking. A full, open, and transparent study of UFOs/UAPs should be carried out. The mission wouldn't be to prove that UFOs were alien spacecraft. That would be assuming the answer. Instead, the task would be to simply understand what the old data could provide and decide how to get new and better data.

As always, the leap from "Oh, that's a freaky story" to "Alien civilizations exist and routinely buzz our aircraft" is so vast that scientists need more than just stories to take it. McDonald's words offer pretty hair-raising narratives, including times that radar picked up UFOs as multiple witnesses were seeing them firsthand. But in the end we are left with stories and nothing more. Still, McDonald was right. Better science is always a good thing. To see how better science might be brought to the subject of UFOs, let's look at those US Navy videos and chart our journey from UFOs to UAPs.

UFOs BECOME UAPs
The modern era begins

The proverbial poop hit the proverbial fan on December 16, 2017. That's when the most eminent, sober, and respectable of mainstream newspapers, the *New York Times*, announced that UFOs were real.

The story, "Glowing Auras and 'Black Money': The Pentagon's Mysterious U.F.O. Program," detailed the existence of new US efforts aimed at figuring out what was behind the mysterious flying things. The program, located squarely in the Department of Defense, seemed to be the first serious attempt to grapple with UFOs since the era of Project Blue Book and the Condon Report.[5] Even more amazing, videos accompanying the article showed Navy pilots tracking strange objects over the ocean. One of the videos included a soundtrack with pilots exclaiming: "My Gosh!" and "There's a whole fleet of them!"[6] Suddenly it seemed that UFOs really were real and, by association, so were aliens. The era of the unidentified aerial phenomena, the government's rebranding of UFOs, had begun.

There is a lot to unpack here, including what exactly was going on with this "mysterious" government UFO program. But let's start with the stuff that's most important and most fun: the actual factual UAPs.

What made the *New York Times* story so compelling were the videos from cameras mounted on the Navy jets. The first of these, from a forward-looking infrared (FLIR) camera, was recorded in 2004 by a pilot from the USS *Nimitz* off the coast of Southern California. The event began when two other Navy jets sighted a white oval-shaped object sitting just above the ocean. As one of the jets drew closer, the object began to rise and followed the jets' own motion before flying off at high speed. Later, another group of jets arrived, and one of these pilots tracked the object on his infrared camera in what became the FLIR video (this pilot never saw the object directly). The second video (and a third released later), called GIMBAL (and GOFAST) were recorded in 2015 off the East Coast of the United States by pilots from the aircraft carrier USS *Theodore Roosevelt*. During a three-week period, multiple Navy pilots reported almost daily sightings of UAPs in this area.

In terms of content, the FLIR video shows a blurry monochromatic image of an oval-shaped object the camera continually attempts to keep within the center of the frame. It's hard to make out what the background is in terms of clouds or sky or sea. There is no audio. The GOFAST video is more compelling; it is sweeping over what is obviously the ocean surface when a small white dot rapidly crosses the screen. The camera then swings its view, attempting to lock on to the object as a voice exclaims, "Wooo! Got it!" The camera continues to track the UAP as it seems to rush just along the wavetops at super high speed. Then there's the GIMBAL video, which shows a white, blurry oblong object that looks like a misshapen Tic Tac. The object is steadily tracked in the center camera frame as the background, which appears to be a cloudscape, sweeps by. Voices, assumed to be the pilots, banter excitedly about how the object is moving against the wind. "Look at that thing, dude," says one voice, as the Tic Tac, which now takes on a decidedly saucer-like cross section, rotates back and forth.

The videos definitely give a hair-on-the-back-of-the-neck feel when you first see them, especially with the pilots clearly delighting in their moment of complete wonder. Just as provocative are later interviews with these pilots. On a *60 Minutes* segment, pilots David Fravor and Alex Anne Dietrich describe their encounter. Fravor tells interviewer Bill Whitaker, "I said, 'Dude, do you, do you see that thing down there?'"

Dietrich adds, "So your mind tries to make sense of it. I'm gonna categorize this as maybe a helicopter or maybe a drone. And when it disappeared. I mean it was just . . ." She never finishes her sentence. It's noteworthy that Dietrich had never spoken publicly about the encounter before. Dietrich tells the interviewer sheepishly, "I never wanted to be on national TV . . . no offense," and then explains why she is coming forward now. "I was in a government aircraft," she says. "I was on the clock, and I feel a responsibility to share what I can."[7]

According to the original *Times* story, these kinds of encounters came to be the principal interest of the Pentagon's Advanced Aerospace Threat Identification Program, or AATIP, which it created in 2007. The program's origin story begins with Nevada Democratic Senator Harry Reid having discussions with longtime donor and billionaire friend Robert Bigelow. Bigelow has a long-standing interest in both UFOs and the paranormal, as well as in commercial space enterprise. Bigelow owns a company that built inflatable space habitats tested on the International Space Station. Reid, with the support of two other senators, sponsored a bill that allocated $12 million to fund a UFO study—a tiny amount by Pentagon standards. From the article it appears that AATIP is the fruit of that bill. Naturally, most of the funds went to Bigelow's company to carry out the study, but AATIP supposedly administered the program. This is how one Luis Elizondo, a former intelligence operative, enters the story.

According to the *Times*, Elizondo ran AATIP. He soon became the face of the media frenzy around UAPs, showing up in a CBS piece, countless newspaper stories, and multiple documentaries. Elizondo ran AATIP until 2012, when the funds dried up, but he claims he continued to work with the Navy and the CIA until he resigned in protest because the government wasn't taking the potentially radical implications of UAPs seriously enough. "Why aren't we spending more time and effort on this issue?" Elizondo asked in his resignation letter. Toward the end of the article comes a description of buildings at Bigelow's facility that had to be modified to store "metal alloys and other materials that Mr. Elizondo and program contractors said they had recovered." If that wasn't enough to make you spit-take your morning coffee, the next paragraph includes a stunning quote from Harold E. Puthoff, an engineer who worked with the program. "We're sort of in the position of what would happen if you gave Leonardo da Vinci a garage-door opener."

A special government program to study UFOs that is housing strange metals recovered from said UFOs is a pretty radical idea, to put it mildly. As you might expect, the Internet, and therefore the world, went crazy. In the wake of the first story, and a number of other pieces that followed, the UFO community felt vindicated. But those of a more skeptical bent were left wanting a whole lot more information—like, for example, where the hell *are* these UAP fragments? When are we going to get to see them? In general, everyone was left with a bag full of fairly stunning questions. Was this the moment when the government would finally reveal what it knows about UFOs, er, UAPs? Most important, are we really about to get proof that aliens exist? It was head-spinning. As usual, however, and as with everything else about UFOs, the fog never seems to clear.

Regarding AATIP and the government UAP programs, things got confusing quickly. In a June 2019 story for *The Intercept*,[8] journalist Keith Kloor quotes a Pentagon spokesman as saying, "Mr. Elizondo had no responsibilities with regard to the AATIP program." Moreover, as Sarah Scoles details in her excellent book *They Are Already Here*, other investigators had a difficult time finding *any* information about AATIP, including what exactly the program's purpose was.[9] Much of this work was done by John Greenewald Jr., who has used the Freedom of Information Act to extensively research government activity on UFOs. Was AATIP more focused on conventional threats, like advanced military hardware from terrestrial adversaries like Russia and China? Or was its main purview really UAPs of the "could be aliens" kind? Greenewald was told by a Pentagon spokesperson, "According to all the official information I have now, when implemented, AATIP did not pursue research into unidentified aerial phenomena." Given the government's long history of using UFOs for public misinformation and misdirection, it's hard to tell where the truth might be. Elizondo, for his part, filed a complaint with the Pentagon claiming a coordinated campaign to discredit him for speaking out.[10]

Of course, aside from the government programs, what really matters are the sightings themselves. What are we to make of trained Navy pilots reporting almost daily contact with stuff that's unidentified and unexplained? What did they and their instruments see, and most important, does any of it present credible evidence for aliens? At this point, we begin a game of "on the one hand . . . but on the other hand" in which believers and skeptics look at the same information and draw different conclusions. For those tending toward belief, the images the cameras captured are hard evidence that these objects are moving in ways that no terrestrial technology could re-create. The objects show no propulsion plumes or any form of "drive," even as they appear to show tremendous rates of acceleration.

Skeptics will point out that many of the effects seen in the three videos can be attributed to the cameras themselves. Mick West, an independent investigator and science writer, has famously connected the most striking aspects of the videos in laboratory tests to the internal motions of the cameras.[11] The sensors on the jets are designed to swivel to maintain a lock on a target. Thus, the camera's own motions, along with the internal optics of the sensors and physical effects like parallax—shifts of view when foreground objects move against distant backgrounds—may explain a lot about what's seen in the videos. Furthermore, skeptics will point out that we are given no context for the videos. They come predigested, already edited by somebody—including the audio, which we have to take on faith is real and was recorded in real time. We also know almost nothing about the specific instruments on the jets and their specific histories. Were they serviced recently? Did they get software updates? The history of science is full of amazing discoveries that went away when sensors were cleaned and calibrated. Given the world-shaking importance of concluding that alien life exists and is visiting the planet, you really need something stronger than edited videos for which you can't get the original data or the instruments that took the data. As compelling, interesting, and super-exciting as the videos and testimony may be, they do not represent the kind of evidence needed to conclude that UAPs equal aliens. Those videos simply don't provide the hard data to tell us if the "objects" are real or just instrument effects. They certainly don't give us the information we need to know if the things are exhibiting motions (i.e., accelerations) that require technologies humans don't possess.

Responding to the kerfuffle they'd helped generate, the federal government moved quickly to release a new and federally mandated report on UFOs[12] shortly after all this came to light. It turned out to be pretty thin, just nine pages. On the one hand, it admitted that the government had cataloged over one hundred sightings it could not explain. On the other hand, it also said that no evidence existed that anything unexplained was extraterrestrial in origin. Once again, depending on your point of view, it was either a big win or just more confusion. In fact, later reporting by the prestigious journal *Science*[13] revealed the chief scientist writing the report was an advocate for paranormal phenomena who had appeared many times on that scientific disaster of a show *Ancient Aliens* (a program I've been invited on a bunch of times but always refused). This left many people wondering just how hard anyone writing the report had *tried* to explain stuff that ended up in the "unexplained" category. They certainly weren't relying on the scientific cream of the crop for expertise.

Along different lines, in the summer of 2023 another UFO media storm hit when David Grusch filed an official "whistleblower" complaint with the US government. In interviews, Grusch first claimed that the US had been collecting crashed UFOs for a while. The idea that the government has artifacts of alien, or at least "nonhuman," technology would be shocking enough. But later Grusch went on to claim that the aliens had engaged in some "malevolent events" in which people may have been hurt. These are pretty extraordinary propositions, to put it mildly. But, once again, no direct evidence was offered. Instead, Grusch's claims were all hearsay. He had never actually seen any of the starships. Instead, he explained that he had talked to people who had talked to people.

If I had to bet, I'd put money down that the Grusch UFO chapter will end up as so many before. Remember *The Estimate of the Situation* we covered when we looked at the previous government reports? That was a secret government report, claimed to exist by a government insider, that concluded aliens were interplanetary. But no copy of the report has even been found. And while some intelligence officials would say that Grusch's claims are legitimate, others have said they are baseless fantasies. So, the old "it's aliens" vs. "oh no it is not" split is being played out within government once again. In the end, I'm doubting that any hard evidence, like a piece of an "alien" technology scientists could analyze, will be available anytime soon. I'd be happy to be wrong, but this does all seem like an episode of *The X-Files*.

Where does this leave us now? Has anything really changed when it comes to UFOs and UAPs? Are we at a turning point in the study of the subject or is this just another cycle of fevered interest followed by more confusion, the kind of thing that has played out many times since 1947? In spite of what may be the usual antics on the part of some folks with axes to grind, I *do* think something has changed. The fact that military pilots are being given the go-ahead to talk about what they see is, I think, a good thing. No matter what UAPs may turn out to be, open transparent reporting is needed, and with that openness comes the possibility of open transparent scientific investigation. That's the only way to ever figure anything out. Some of these investigations are beginning as I write these words, including a NASA panel on the subject. When the NASA panel held its first public meeting in the summer of 2023, they reported an analysis of one of the three famous Navy UAP videos. Using some basic methods in geometry, they determined that the object seemingly skimming rapidly above water in the GOFAST video was actually at 13,000 feet and was moving at about forty miles per hour (which was the wind speed at the time). There is nothing very extraterrestrial about forty miles per hour. I can go almost that fast rolling downhill on my mountain bike and that's without pedaling. Even more to the point, the panel said that only 2 to 5 percent of the reported UAP sightings were deemed "possibly really anonymous." These conclusions show how even a little science can go a long way. If done well, the NASA panel and other efforts could provide a master class in how science goes about its remarkable business of unbiased exploration. In a world full of science denial of many forms, that in itself would be a great service.

So now we come to the next big question: If this new wave of Navy UAP sightings doesn't give us what we need to know to address the alien question, what kind of data do we need and how do we get it?

HOW TO GET REAL ABOUT UFOs
What a true scientific study would look like

The most important difference between this cycle of UFO interest (or hype depending on how you look at it) and those that came before is the government's involvement. From congressional mandates to Pentagon programs (whatever their real intentions are), this time around it appears the government agencies are willing to be somewhat more honest by admitting that they're seeing something and they don't know what it is. That's important for one very important reason: money.

Any attempt to bring real science to the understanding of UAPs will require funding at a scale that the private sector won't marshal. Assuming that the government is willing to put some funding into an open, fully transparent scientific exploration of the subject, what would that look like? It's a great question because there are already some groups trying to feel their way through the entry points into exactly this kind of research.

The first and most important thing you must know, however, is what this kind of research *won't* be doing. It won't be trying to prove that UFOs are driven by aliens. Finding that UFOs are alien SUVs would be just one possible end point of a project that starts off *agnostic* about its subject. Good science begins without any preconceptions.

That perspective is very different from "See this stuff? We don't know what it is but we are really hoping it's Viking Elves from Dimension 9, so go find evidence to prove we're right." As they say in the science-fiction classic *Dune*, "A beginning is a very delicate time." You have to start a project like this knowing exactly what the mission is. Proving your favorite hypothesis in not the mission. Going wherever the data leads—that's the mission.

So, fine, we have these UAPs that keep zipping around our aircraft. How do we figure out what they might be? Ironically, what's needed in a scientific search concerning UAPs is not so different from what's required of my colleagues and me in our scientific search for biosignatures and technosignatures as we look for life on distant worlds. What's needed is *hard data* from *verified instruments* collected via a *rational search strategy*. Let's start in the middle of that sentence and focus on the idea of verified instruments.

If and when my colleagues report the detection of, say, a biosignature on an alien world, they'll do so using a light detector that's been hung on a telescope. That entire apparatus constitutes a scientific instrument. That instrument will have had every aspect of its behavior explicitly and exactly characterized. Astronomers will know how well it responds to long-wavelength infrared light and how well it responds to shorter-wavelength visible light. They will know how the instrument behaves at 40° Fahrenheit and how that behavior changes at 60°. The list of these characterizations for the instrument will be exhaustively long and very boring, but if you're going to claim that your instrument saw something amazing, you'd better know everything about that hardware six ways from Sunday.

If we're going to start a UAP research program, we're going to need a bunch of well-characterized instruments. But what kind? That depends on what kind of data we want. We are certainly going to want imagers (cameras). We'll also want to take images in as many wavelengths as possible. Visible light is good, because that's where our eyes work and the atmosphere is quite transparent to it. Infrared light is good for heat signatures like propulsion or other thermal sources. Beyond imaging, we might also look for radio waves being emitted from the objects, because radio is good for both communications and sensing (radar uses radio wavelengths). We will also want highly accurate information about motion, to track speeds, and, more important, changes in speed (accelerations and decelerations). These tell us about forces, which could indicate propulsion technology. Whatever we choose, we'll have a suite of instruments that are ready to be sent out to collect our hard data.

Now for the last part: the rational search strategy. We have three choices. We can look up. We can look down. We can look all around. Yeah, I know that sounds stupid, and you may be wondering what fool gave me that PhD, but hear the argument out. Looking up means deploying our instruments as ground stations. We could try putting upward-looking cameras and radar units in a grid throughout large swaths of the country, or we could concentrate them in places that seem to have more UFO activity, like military bases or parts of the oceans where naval exercises are held. They would then sit there taking data all the time (expensive) or when triggered by some kind of alert. We could also put up a network of satellites looking down on the Earth with various detectors. They, too, could be continuously monitoring or waiting for triggers. Finally, we could equip lots of planes (private and military) with sensors built for the search. That's the looking-around strategy. These sensors, once activated, perhaps by pilots who see something from the cockpit, could even be the triggers that activate the ground and satellite systems.

Plans for such systems are already being discussed. The Galileo Project, led by Harvard astronomer Avi Loeb, is already building smaller-scale versions of the upward-looking ground stations, taking cues from meteor scientists, who already have a version of ground-based detection systems. In addition, NASA has commissioned a panel of scientists and engineers to look at how their existing network of satellites might help. NASA already studies Earth quite a lot for subjects like climate change.

What happens once we have all these petabytes of images and velocity information? The answer is both simple and complicated. We look for something weird. That means looking for, say, accelerations of an object that's been detected in multiple-wavelength images (so we know it's real) beyond anything a known engine could produce. The highest accelerations obtainable by an air-to-air missile is on the order of a few hundred G's. Are the detected UAPs accelerating faster than that? Better than just seeing something accelerate wildly in a straight line would be watching it turn so quickly that it would shred any known metal humans know how to produce. That would be pretty awesome and would indicate something weird is at play. We could also look for hovering motions with no propulsion signatures. If a well-characterized object (again seen at multiple wavelengths) is observed just hanging in place, that would definitely trigger our weirdnometers.

Making sure these detections aren't artifacts of the instruments, the data-collection methods and the data-analysis algorithms will be the complicated parts. Remember that we may be collecting huge amounts of data. Sorting through it all to find the few things that constitute a signal above the noise will be its own huge task, but as they say on *The Mandalorian* (and who doesn't love Baby Yoda?), "This is the way."

This is the way a real, scientifically valid research program aimed at getting to the root of the UFO/UAP mystery gets carried out. As to the pilots of such craft, that's a whole other story.

CHAPTER 4

What If They Are Aliens?

IF UFOs ARE ET, HOW'D THEY GET HERE, AND WHAT THE HELL ARE THEY DOING?

Aliens are not actually magic. That includes whatever turbocharged hovering, hypersonic spacecraft they might be flying. Whatever aliens do, no matter how miraculous or freaky it may seem to us, it's still going to be based on physics, chemistry, biology, and so on. The physics, chemistry, and biology the aliens are working under may be totally different from anything we've imagined. It may be so far ahead of our understanding that in comparison we look like amoebas trying to do calculus. But, at the core, whatever aliens do will be based on science. The cosmos we live in is the same cosmos they live in. That shared universe is full of patterns, linkages, and networks of cause and effect. Science, no matter who does it, is simply about digging into those patterns and coming back with some understanding of how they work. If you want to get serious about aliens, you gotta think hard about aliens . . . and physics. That's when the real fun begins.

In this chapter, we're going to take seriously the idea that a species with advanced technology can visit Earth. Under that premise,

WHAT IF THEY ARE ALIENS?

let's ask: How might the technology they're using work? That question will take us into territory at the edge of modern physics. Einstein's relativity, quantum mechanics, stellar astrophysics—getting a handle on the tech of advanced civilizations forces us into these topics. The view of the universe that these domains of physics offer is stunning. It's going to be worth the hike up their mountainous terrain. Getting even a glimpse of these biggest science ideas—things like space warping and extra dimensions—will give us enough purchase to also see what might be possible for aliens from distant worlds showing off on their trips here. It also helps us see where human civilization might be heading if we last long enough.

Let's begin at the beginning and ask the most basic question about how anyone can cross the insane, vast, mind-emptying distances between the stars.

INTERSTELLAR TRAVEL[1]
If UFOs were aliens, how did they get here?

I'm going to give you the bad news first. The distances between stars are so large that they might be impossible to routinely cross. Sure, maybe you send robot probes that reach their target in two hundred years (and then you need another century or so for a message to get back). But the possibility that you, I, or anybody else can pop around to the best vacation planets in the galactic empire may simply be excluded by the laws of physics.

Or maybe not.

This is the kind of landscape we have to deal with when we try to navigate the question of aliens and interstellar travel. We absolutely, positively know the distances between the stars. We also know for certain that the universe imposes a speed limit when it comes to crossing those distances. What we need to do next is imagine, based on what we know about the structure of reality, how aliens might get around those limits.

If UFOs *are* spaceships from other star systems, then how might they (or us in the future) cross the great interstellar voids?

The first thing we need to address this question is an understanding of just how big, big, biggity-big space really is (to paraphrase the great Douglas Adams). Yeah, I know you *think* you know how big space is, but trust me, it's bigger. Every time I have to deal with these kinds of distances in my research, my capacity for freaking out at the scale of the cosmos (even our wee corner of it) is entirely and forcefully renewed.

Astronomers measure interstellar distances in light-years, which, I know, is confusing. A light-year is the distance light travels in a year and spans 6 trillion miles. That's a six with twelve zeros after it—6,000,000,000,000 miles. You have probably walked a mile and driven thousands of miles. All this takes you through only the first three zeros. The other nine require a heroic feat of imagination. If you're looking for a familiar comparison, it's the same as traveling all the way around the Earth hundreds of millions of times. Imagine how many connecting flights and pointless waits getting stranded in O'Hare Airport that would imply.

Another way to understand a light-year is to consider the distance from the Sun to the edge of the solar system. If the Milky Way galaxy is our local city of stars, then the solar system is the house we were born in, and Earth is one room in that house. In 2006 we launched the fastest space probe ever developed, New Horizons, and sent it to Pluto, which can stand in for the edge of the solar system. The distance to Pluto is about one thousand times shorter than a light-year. Our solar system, where all human activity on planets and in space has played out, is a tiny fraction of a light-year across. And here's the real point to ponder: even though New Horizons was hurtling through space at 36,000 miles per hour, *it still took about ten years to reach Pluto*. From that factoid, we can conclude that it would take New Horizons at least twenty thousand years to travel a single lightyear.

WHAT IF THEY ARE ALIENS?

That's a very long time, but it still doesn't even get us all the way to interstellar distances. There's nothing much out there at one light-year away. The Oort Cloud, where most of the solar system's comets live in cold storage, extends out to around a light-year, so even out at this distance, you're still technically *in* the solar system.

You have to travel around three more light-years to reach the nearest star, Alpha Centauri. A journey by New Horizons to that star would take around eighty thousand years, and most stars are way, way farther away than Alpha Centauri. The Milky Way galaxy is about a hundred thousand light-years across. Even if we stay in our local neighborhood, the distance to the nearest interstellar Starbucks has to be measured in hundreds or thousands of light-years. That would be tens of millions of years of travel time for our fastest space probes.

All of this serves to confirm that, yes, space is frackin' big. If UFOs really are interstellar visitors, then these are distances they must routinely cross. These are also the distances we must learn to cross if we are to become an interstellar species, aliens to someone else.

Any attempt to traverse those distances runs into a fundamental fact about the universe. Nothing can travel faster than light speed. This is not just a fact about light. It's a fact about the very nature of physical reality. It's hardwired into physics. The universe has a maximum speed limit, and light just happens to be the thing that travels at it. Actually, any particle without mass (like light) travels at light speed, but nothing anywhere can travel faster. This speed limit is so fundamental that it's baked into the existence of cause and effect. The finite speed of light is what *forces* effects to come after causes like dishes shattering only after they get knocked off tables.

There may, of course, be more physics that we don't know about that's relevant to this question of interstellar travel. Still, this speed-of-light thing is so important to all known physics that if you want UFOs to be spaceships, you can't get around it by just saying, "Oh, they'll figure it out." You gotta work harder than that.

Now let's dig into the problem. Given these insane interstellar distances, how can we extrapolate from the physics we *do* understand to see how those aliens (or us in the future) might cross the cosmic void. We have a few possibilities.

Generation Ships: *Depending on their biology, the life span of our hypothetical aliens might be shorter than the centuries-long journey required for slow sub-light-speed travel between stars. This is certainly the case for us. If you are awake the whole trip, you'll be dead by the time you get there One way around this dead-on-arrival dilemma is to have children along the way. You'll still be dead, but your kids or grandkids or great-great-grandkids' offspring will make it. Generation ships (also called century ships) are one way that interstellar travel might be possible. Those ships would have to be pretty big, though, to carry an entire colony of space travelers. It would be hard to miss one of these if it pulled into orbit. Also, you might imagine that those grandkids would be pretty pissed off about having to spend their entire lives on a smelly space cruiser. Maybe the kids are the ones erratically flying the UFOs; that could explain a lot .*

Cryosleep: *Another obvious answer to the dead-on-arrival dilemma is to hibernate. Cryosleep technology would basically "freeze" the body's metabolism (or at least slow it way down) for the duration of the journey. In spite of being a staple of science fiction, no one has come close to getting this to work for higher animals like mammals. Still, it is the kind of solution that doesn't require new physics to magically exist (maybe just new biology). Also, if "post-biological" machine-based life is really a thing (as we'll explore later), then maybe some aliens switch to silicon-based digital form and this question of long timescales is not even an issue anymore.*

Light sails: *While no one has ever been blown down the street by a ray of sunlight, photons (light particles) do exert a force—a push—on matter. If you could extend a large enough sheet of material in space, you could use the Sun to propel you through space. The idea of such solar sails has been around for a long time, but in 2016 Philip Lubin of the University*

WHAT IF THEY ARE ALIENS?

of California–Santa Barbara[2] proposed using very powerful giant lasers rather than the Sun, to provide the light for interstellar sailing. With a large enough ground-based laser, you could accelerate a sail in space (and a ship tethered to it), up to nearly the speed of light. This way you could cross the distance between nearby stars in years or decades rather than thousands of centuries. The billionaire philanthropist Yuri Milner was so taken with this idea that he gave $100 million to development of a project called Breakthrough Starshot.[3] It's a long-term effort with a thirty-year timeline, because that's how hard the technology will be to develop. The hitch for UFOs using this tech is that you need another giant laser in the target star system to slow you down if you want to stop and visit.

Wormholes: *If the speed of light limits how fast you can travel through space, then the best solution for interstellar travel might be avoiding the through part of the problem. That possibility was one of the gifts Einstein gave us with his general theory of relativity (GR). In relativity, space is not an empty void. Merged with time into a single entity called space-time, it is like a flexible fabric that can be bent, stretched, and folded. Wormholes are a kind of space-time tunnel that uses this folding to join two widely separated regions of space together. While such wormholes (aka Einstein-Rosen bridges) are most definitely allowed in GR, they are unfortunately unstable. Once a wormhole is formed (by whatever means, natural or otherwise), it will almost instantly slam closed. If aliens wanted to use wormholes to build a kind of galactic transit system, they would need to find something physicists call exotic matter. This is stuff that has true antigravity properties, that literally pushes space apart. If aliens had and could control exotic matter, they could force the two mouths of a wormhole open and connect two distant parts of a galaxy. Before you get too excited, there's a big hitch here. Exotic matter isn't real. It's just a term you can add to the GR equations. Include that term into formulas, and it changes how they behave. Hooray! But that doesn't mean the term represents anything that actually exists in the universe. Still, that antigravity term is possible within the framework of relativity's physics equations, so if* exotic matter is more than a physicist's pipe dream, maybe it could serve as the means for fast interstellar travel.

Warp Drives (aka Hyperdrives): *Warp drives—or hyperdrives or frameshift drives or whatever you want to call them—are a staple of science fiction. If you want your characters to easily travel around the galaxy, just put a warp drive on their ship, and no one will ask any questions. If aliens could build a warp drive, they would once again be using the "fabric-of-space" idea from Einstein's GR. The drive doesn't push you through space from one place to another. Instead, it creates a "warp bubble" that stretches and then relaxes the space-time around you. You don't travel through space faster than light; you warp and unwarp space itself faster than light. While nothing can travel faster than the speed of light through space, space (i.e., space-time) can move at whatever speed it likes. The nice thing about warp bubbles is that, like wormholes, they are also theoretically possible in GR, as physicist Miguel Alcubierre showed in a famous 1994 paper.* But there are, as you might expect, some really big problems with warp drives. Once again, you would need that nonexistent exotic matter. Even more problematic is that warp bubbles may generate huge shock waves of high-energy gamma rays as they move along. Once you dropped out of warp, this blast of energy would fry everything in your path and sterilize any planet you were visiting. Not the best way to announce your arrival at a starport.*

Quantum Mechanics: *Quantum physics, our über-powerful theory of the atomic and subatomic world, is notoriously weird. In quantum-mechanics physicists are forced to talk about particles being in two places at the same time or two particles instantly affecting each other even though they're on opposite sides of the universe. Even a hundred years after quantum mechanics became the most accurate, potent physical theory ever created, the basis for all our electronic miracles, we still can't say we understand what it's telling us about reality. Personally, I think that's pretty cool. What it*

---

* Miguel Alcubierre, "The Warp Drive: Hyper-Fast Travel within General Relativity," *Classical and Quantum Gravity* 11, no. 5 (1994): L73–L77.

means for interstellar travel, however, is that there might be something hiding in quantum mechanics that allows you to bypass GR's apparent restrictions about space and time. Some folks working on merging quantum mechanics with GR into a theory of quantum gravity even believe that space-time may not be fundamental. Instead, it might emerge out of some deeper aspect of quantum reality. So, yeah, quantum mechanics could have some tricks up its sleeve that a sufficiently advanced alien species might know about and exploit for interstellar travel. But be careful. Unlike the other items on our list, here we are certainly pulling hope out of our keisters. There is no physics here yet, other than waving at the passing quantum weirdness.

So that's it. As far as we know (which may not be far enough) that's all we or aliens have when it comes to physics and interstellar travel. Now, a good science-fiction writer might find other creative ways to imagine getting from one star to the next, but the list above pretty much exhausts what a scientist would propose based on what we know about reality (which is a lot). The important thing to know is that, in terms of experimentally validated physics, after the first two possibilities, Elvis has most definitely left the building.

To complete the picture, we can also ask how an alien interstellar drive technology might affect what they're doing here. If they're restricted to moving just at or below light-speed, that would limit their ability to build a galactic civilization. When it takes two hundred years to send a diplomat between two planets and no one lives more than a hundred years, you have a problem. Yes, maybe you could solve that problem with cryosleep, but it would still take two hundred years for that diplomat to arrive. Much will have changed on the home planet during that time. Will there even be the same kind of government on either world by the time the diplomat gets there? And after the negotiations, it will still take another two centuries to deliver the answer. That's a long time to wait for "They said, 'Go to hell'" (or whatever).

All of this means that, if warp drives or other faster-than-light tech is *not* possible, then our ideas about interstellar civilizations may be very wrong. If no one can travel faster than light, maybe it's every solar system for itself. In that case, you never get galactic empires, just individual planetary cultures. These cultures might send settlement missions out to cross the stars once in a while, but given the distances and travel times, even if those settlements succeed, they'd quickly diverge culturally from the home world. If this is what happens, then aliens visiting Earth are not representatives from some Zorgovian Galactic Federation with vast experience of many worlds and many cultures. Instead, they'd be one-offs, and we might be their first visit anywhere.

Now let's stop and take a deep breath. Everything I've just spun out for you is a deep pile of speculation. It does, however, come from taking the science seriously. That's what makes it so much fun to explore. Life always exists within constraints that the universe imposes on it. Technology can press against those constraints, but it won't make the fact of the constraints go away. Engines need power. Machines break down. Any alien technology, no matter how awesome, will have to deal with those constraints, so now let's ask what other kinds of super-advanced tech visiting UFO aliens might have in their toolkits.

ALIEN TECHNOLOGY
Inside Luke Skywalker's garage

UFOs are supposed to do lots of "holy crap" things that defy the laws of physics. They appear out of nowhere and disappear in a flash. They travel at hypersonic speeds and stop on a dime. They make sharp turns at impossible angles. These behaviors, if real, might defy *our* understanding of the laws of physics, but that doesn't mean they're defying physics itself. As I've been emphasizing until you want to smack me, the physical world is based on physical laws even if we don't know them all. If alien spaceships exist and can do the things UFOs are reportedly able to do, what unknown physics might be behind their technology?

WHAT IF THEY ARE ALIENS?

Let's start with the disappearing thing. Lots of UFO reports have objects appearing on radar or to witnesses only to vanish quickly thereafter. If this were real, it might imply some kind of "cloaking" capacity like what the Romulans have in *Star Trek*. The ships would really be there, but they could become invisible anytime they choose. How would that work?

As discussed before, all light is electromagnetic radiation (waves) of various wavelengths. Radio light (radio waves) comes in long wavelengths the size of buildings, while X-rays have wavelengths the size of atoms. That means, to make something invisible, you must do something to those waves. But what does "invisible" really mean? Seeing something like a spaceship with light involves two steps. First, you need to bounce electromagnetic waves off the ship, and second, you need a detector to pick up the reflected waves. The detector could be your eyes or it could be a radar dish. The easiest way to mess with this seeing process would be to make the ship from a material that simply absorbed the incoming electromagnetic waves, leaving no reflection for the detector to detect. Or you could develop materials that simply bend the light around the spacecraft, so it never makes contact with the ship at all and has no need to be absorbed. Finally, you could also have the spacecraft emit light that mimics its surroundings so that it always blends in perfectly. This would be like an adaptive, intelligent camouflage.

Aliens possessing any version of this kind of tech would not be a huge stretch scientifically. We humans are already at the edge of some of these capacities. Over the last few years, physicists have developed "metamaterials," which have the capacity to redirect and bend incoming light. By covering an object with a thin layer of metamaterial, you could deflect incoming light along the craft's skin until it gets ejected on the other side. Light from the other side would also bend around the craft. This way an observer (you) would always see whatever was behind the craft, making it invisible. We're still a long way from building a tank or fighter jet from metamaterials, but the basic principles are being tested right now in terrestrial laboratories.

Hovering or making right hand turns at Mach 50 is another story. Science fiction is full of antigravity technology. Think of Luke Skywalker's speeder skimming just above the ground, directly canceling the force of gravity. To understand how the force of gravity might be *turned off*, we first need to understand exactly what physicists mean when they talk about forces. First, a force is at work when some stuff (matter) pushes or pulls on some other stuff. After centuries of study, physicists have found only four basic forces: gravity, the electromagnetic force, the strong nuclear force, and the weak nuclear force.

Everyone is familiar with gravity, the thing that keeps planets in orbit around stars and your feet pinned to Earth's surface. Electromagnetism is the force between electrically charged particles like electrons and protons. It's basically the force behind all of life, because chemical bonds are really electromagnetic bonds. It's also what's behind all our modern digital technology. The strong nuclear force is what keeps atomic nuclei together. It's the basis for the existence of the different elements. Finally, the weak nuclear force is involved in the nuclear reactions that happen in stars. There's a lot more to this story, but this thumbnail description will work for our purposes.

Everything that happens in the universe, no matter how complicated, comes down to just these four ways to push or pull. That fact is remarkable by itself but still doesn't tell us exactly what's doing the pushing and the pulling. The answer is: force particles.

In the 1940s physicists began to realize that each push-and-pull between subatomic particles (like electrons and protons) always involves the exchange of a force particle. The technical name for these special particles is *force bosons*. When two electrons, which both have a negative electric charge, electromagnetically repel each other, it's not because they actually bounce off each other. What's really happening is that they are swapping one kind of electromagnetic force particle, a *photon* (a particle of light). Photons are the electromagnetic force's force boson. Each of the other forces has its own force particle too. The force boson for gravity is called a *graviton*. Gravitons have never

WHAT IF THEY ARE ALIENS?

been observed, because we don't have experimental machines powerful enough to probe them, but physicists think they have to exist.

We need to understand all this because technologies for manipulating a force come down, in one way or another, to technologies for manipulating the associated force boson. We have computers and cell phones and all the other wonders of electronics because we're masters of manipulating electromagnetism's force boson, the photon. We can make photons dance to almost whatever tune we want. The other forces and their bosons are another story entirely. Direct control of them would require insane amounts of energy. That leaves us with only ham-fisted control over them. When it comes to using gravity in technology, for example, we're not doing much better than cavemen dropping rocks on each other's heads. Even thinking about how to control gravity at the graviton scale is beyond us. But maybe aliens are better at working with gravitons than we are.

Creating antigravity devices could require gaining control of gravitons in ways we can only dream of. Maybe aliens know how to redirect them on the sub-sub-subatomic scale. Maybe they know how to reduce the graviton exchange between Luke Skywalker's speeder and the ground on his planet, Tatooine. In that case, the speeder (or a UFO) could hover above the ground as if the gravitational pull of Tatooine were completely absent. Commanding gravitons might also be the way to create artificial gravity in a spaceship, as on *Star Trek*'s *Enterprise*. How else could everyone stand around on the bridge as if they weren't in the weightlessness of space?

Having direct control over gravitons could also help UFOs decelerate and accelerate so rapidly in those hypervelocity turns. When your car makes a turn, the road is exerting a force on the tires that compels the rest of the car to change direction. If you don't have your seat belt on, you'll feel yourself sliding along the seat in the direction opposite the turn. It's the seat belt that communicates the force that forces your body to take the turn with the rest of the car. The resistance of your body to making that turn is called *inertia*.

If you tried making an instantaneous sharp turn traveling at 3,000 miles per hour in a spaceship, inertia would force your body to smoosh through the seat belt and splatter against the wall.

Now imagine you were going into one of those hypervelocity right-hand turns but also had direct control of gravitons. You could, perhaps, create a spray of gravitons at just the right time and in just the right way to create a gravity field that cancels the inertia of your body. That way, instead of having the seat belt catastrophically force you into the turn (and turn you into jelly), the fake gravitational field could perfectly balance everything out, allowing you to change directions without any drastic tugs or pulls.

Is any of this possible? Who knows? What I've just done is written a science-fiction story wildly extrapolating physics we do understand to create a semi-plausible possibility. It may be that none of it is possible. Maybe the physics just can't work that way. For one thing, to create some of the technologies I just described might demand energies found only near a black hole or just after the Big Bang. If aliens can build graviton manipulators, they'd have to have easy, repeatable access to the same level of power at work in the universe's most extreme objects and conditions. That's kind of a tall order, but these are the kinds of things you have to deal with if you really think UFOs are aliens whipping around our airspace. Maybe they *have* developed those kinds of capacities. God knows I'd be showing off if I had them.

INTERDIMENSIONAL ALIENS
Hey, man, get off my plane

There is another possibility entirely. Maybe UFOs aren't traveling through space at all. While there's a long history of people claiming UFOs are aliens from other star systems, there's also a long history of people claiming they're from other *dimensions*.

"Extra" or "other" dimensions is not an idea specific to the UFO

WHAT IF THEY ARE ALIENS?

crowd; scientists have been obsessed with the question forever. Could there be a connection between science's other dimensions and those of UFO enthusiasts? For an answer, we're going for a short dive into one of the most mind-stretching parts of mathematics and physics. If we want to know if aliens come from other dimensions, we first need to ask, what the hell is a dimension, anyway? This, as the Canadians say, is "big fun." I can't tell you how much I love the idea of other dimensions (well, actually, I'm about to tell you in great detail). I love other dimensions, not for their relation to aliens but for their relation to physics. We'll get to the aliens in a minute.

Talking about dimensions really means talking about capital-*s* Space. How much of space is there, and how much can we access? To answer this, let's define *dimensions* in a way that brings the idea down to Earth.

You live in three dimensions. What does that look like? Well, if you want to, you can walk forward or backward. You can also move left or right. You can also move up or down.

And that's all she wrote.

Other than forward/backward, left/right, and up/down, there are no other independent ways to move through space. What a physicist means by "independent" is that any and every movement you can possibly make will be some combination of forward/backward, left/right, and up/down. You literally run out of space after these three choices, and that's what tells you that space is three dimensional. All the stuff you see out in the world, like spheres or cubes or pyramids or whatever—it all "lives" in three dimensions. No more. No less. This idea by itself is pretty cool. But it gets cooler.

Near the end of the nineteenth century, mathematicians began theorizing about dimensions beyond the standard three. They started asking themselves about the properties of four-dimensional spheres, seven-dimensional cubes, and ten-dimensional pyramids. They didn't think these extra dimensions were physically real, but the tools of mathematics gave them access to these abstract spaces like explorers at the edge of a new continent.

The phrase "tools of mathematics" sounds pretty abstract and imposing. When I was eighteen, I thought it was a good thing to say to impress girls (mostly it didn't). Using those tools just means doing a calculation to figure something out—like the volume of a sphere, how much space there is inside it. For a three-dimensional sphere, the volume is proportional to the radius cubed (R^3). If you want, you could also imagine that sphere exists in a space where there's one more dimension (one more direction to travel in) and redo the same calculation. This time you'll find that the volume is proportional to the radius raised to the power of 4 (R^4). That extra power of the radius tells you there is literally more space inside the four-dimensional sphere than the three-dimensional one. The extra dimension means more space. But the main point here is simpler. To do any calculation in higher dimensions, just repeat what you did in three dimensions, but add another step for each extra dimension. It can be more complicated than this, but you get the idea.

Even with the math, it can be hard to imagine what a four-dimensional sphere looks like, but taking a cue from the mathematics we just described, we can bootstrap our way to it. It's pretty bracing to even get a glimpse of these hyperdimensional possibilities, so let's give it a try. The idea is to start with lower dimensional worlds (fewer than three) and work your way up. We'll start by imagining what a sphere looks like in 1D. Reality in a purely 1D world would be just a line stretching infinitely in the left/right direction. That's all the space there could be. A sphere in 1D would just be a line as long as the sphere's diameter (twice its radius).

Now let's move to a 2D sphere, which is easier to imagine. First of all, what would a 2D reality look like? It would be a plane, like a sheet of paper, that extends to infinity in all directions. A sphere in 2D would just be a circle, like one you'd draw on that sheet of paper. Finally, we all know what a 3D sphere is, because we have lots of examples: beachballs, baseballs, bowling balls.

Line, circle, baseball: they are all spheres but in different dimensions. Now let's go higher.

WHAT IF THEY ARE ALIENS?

The important thing for going to 4D is seeing how a 3D sphere connects to a 2D sphere connects to a 1D sphere. Let's imagine what would happen if a 3D sphere (say, a baseball) "passed through" a 2D world. Before the baseball and the infinite plane comprising the 2D world touched, 2D creatures living on the 2D plane would have no idea anything was about to happen. As soon as the baseball made contact with the plane, however, a point would suddenly appear that the 2D creatures could "see." As the baseball pushed through the 2D plane, that point would then become a small filled-in circle—the intersection between the baseball and the 2D plane. The circle would grow until the baseball was halfway through the plane, and then it would shrink until it became just a point again. When the baseball lost contact with the plane, the point would disappear. Imagine this like a baseball rising to break the surface of a completely still body of water until it was clear and in the air. The baseball's "imprint" on the surface is the connection between the 3D world and the 2D world.

Now we can play the same game going from 4D to 3D. This is not easy, but keep going back to the analogy of the baseball and the water surface; that should help. If a 4D sphere (we will call it a hypersphere) passed through our 3D world, what would you see? Like that baseball first touching the water's surface and producing a little circle, you'd first see a tiny ball magically appear in front of you from nowhere. That would be the 4D hypersphere first touching our 3D "plane" of existence, and it would startle the hell out of you. Totally freaked out, you'd watch as that little ball grew bigger and bigger as the hypersphere pushed through our reality. The ball would stop growing only when the hypersphere was halfway done with its crossing through our world. Next, the ball would start shrinking again until—poof!—it disappeared when the hypersphere completed its crossing of our 3D plane. At this point, a stiff drink or a call to your therapist would be in order.

This kind of thought experiment demonstrates the incredible power of abstract mathematical thinking (exactly what you were just doing).

A BRIEF GUIDE TO ALIENS

The utility of hyperdimensional mathematics became very clear to physicists around 1900. By taking complex problems from our three-dimensional world and "projecting" them into higher-dimensional spaces, new insights were gained, and venerable old problems got solved. The most famous example of adding dimensions to a problem was Einstein's theory of relativity. In relativity, time is a fourth dimension alongside the three dimensions of our familiar space (forward/backward, left/right, up/down). Relativity and its 4D space-time was the greatest revolution in physics since Newton.

The spectacular success of mathematical physics in using extra dimensions (i.e., relativity) was widely reported in newspapers of the early twentieth century. That's how the idea of dimensions as other "planes of being" caught hold in the popular imagination. The spiritualist movement, famous around this time for its séances, picked up on what was happening in science and began applying this new language to its adherents' beliefs, giving them a sheen of scientific credibility. For spiritualists, the higher dimensions were the domain of the dead. That's where they "lived," and that's where they could be contacted by bridging the dimensions. Soon extra dimensions and parallel worlds came to be associated with a lot of paranormal beliefs.

Once widespread UFO sightings began in the 1940s, some folks quickly made the connection to higher dimensions. It was the French scientist Jacques Vallée who really popularized the idea. Vallée is a very interesting and very non-mainstream character who worked on mainstream technical problems like early human-computer interfaces. Vallée also has an enduring interest in UFOs. In the late 1970s, he became a student of J. Allen Hynek, an astronomer who was part of the Condon Committee. Hynek later became a convert to the extraterrestrial hypothesis, and he coined the close-encounters classification system for describing how close a witness gets to a UFO alien.

Vallée, however, went far beyond Hynek. Drawing on the psychoanalyst Carl Jung's writings, Vallée became convinced that UFOs were not traveling to Earth from alien planets via three-dimensional space. Instead, they were coming from other dimensions. This became the "interdimensional hypothesis." Like those 4D spheres passing through 3D space, UFOs were crossing between dimensions to visit us. This, for Vallée, explains why UFOs could appear and disappear both visually and on radar.

Since Vallée's work, the interdimensional hypothesis has maintained a steady, if small, group of proponents. The extraterrestrial hypothesis is definitely the majority perspective, but the idea that UFOs represent beings from other dimensions retains its attraction. This is particularly true for folks who tilt toward belief in the paranormal.

At this point, I could continue telling you how remarkable, how extraordinary, and how beautiful the use of hyperdimensional thinking is in modern physics. I have spent some of my most compelling moments as a theoretical physicist wandering around six-dimensional donuts, letting the details of their shapes reveal the boundaries between order and chaos in, for example, the motion of planets in the solar system. But the thing about extra dimensions is this: they don't really exist.

From the physics point of view, there has never, ever been any evidence of a single extra dimension of space beyond the three that we experience. And it's not for lack of trying.

Let's look at Einstein's four dimensions of space-time. What really happens in Einstein's theory of relativity is that you take time and "spatialize" it, meaning you treat it mathematically like another dimension of space. However, time is always different from the other dimensions. To see this, let's point out the obvious fact that I can go back and forth in each of the three spatial dimensions (left, then right; forward, then backward), but the sad truth is, I can go only one way in time. It's sad because that giant one way sign on Time Street is why we die. In spite of all the great science-fiction stories about time travel, there is zero evidence that's possible. Time is not space, and it's not an extra dimension of space.

On the other hand, it is true that there have been some theories in physics positing extra space dimensions. String theory is a famous example, which claims that the universe really has ten spatial dimensions. In string theory, we can't see the seven extra space dimensions because they're curled up tight, kind of like the way you roll up a piece of paper to make a straw. It's a cool idea but, in spite of decades of effort, string theory has never produced anything close to an experimentally verified prediction, and it just doesn't look that promising anymore to the majority of the physics community.

There's also the multiverse idea. This is the theory that the Big Bang didn't produce just one universe but rather produced many "pocket universes" like our own. The multiverse includes infinite versions of you doing infinite versions of whatever you are doing right now. Unfortunately, these kinds of parallel universes won't work for UFOs, because they aren't really parallel. They don't exist in other dimensions. They're all part of the same space of the Big Bang and, by definition, are so far away that UFOs from those other universes could never ever reach us. Finally, while one version or another of the multiverse idea is great for science-fiction writers (I love you, Marvel Cinematic Universe), there has never been any evidence supporting it. It's just an idea that a handful of physicists favor, while the rest of us are either "meh" about it or actively think it's very wrong.

So, I am sorry. Really. I love the idea of other dimensions in mathematical physics, and I love them in science fiction. But there is zippo evidence that the number of *real* dimensions of *real* space in the *real* universe is more than three.

BUT WHAT ARE THEY DOING HERE?
The high-beam argument and some other questions

We've seen the kinds of technology aliens would need in order to reach us across interstellar distances. We've also seen what tech

WHAT IF THEY ARE ALIENS?

they'd need for their jacked-up UFOs to behave in the impossible ways some folks claim they do. Finally, we also poked our heads into the extra dimensions where yet others claim UFOs come from. We now have a good handle on the what, where, and how of UFOs and aliens. Before we leave the subject and go on to the scientific search for aliens in space and not on Earth, a couple of important points still need to be addressed. I want to touch on these things because, as a scientist who thinks about aliens every day, some common themes about UFOs color much of the debate for me.

The first is what I call the high-beam argument. A lot of the debate around UFOs centers on seeing objects in the sky doing amazing things. But we never seem to be able to get a really good look at them. "I saw these lights in the sky" is the often-quoted refrain. The lights appear, they move in strange and impossible ways, then they fly away at impossible speeds. Stuff is seen in the day too, but it always disappears before high-resolution images can be taken. The gist of this is simple. The aliens never just announce themselves and stand still so everyone can get a good look at them. They don't want us to know they are here. They want to stay hidden. This aspect of UFOs is what leads to my big question: If they want to hide, why do they suck so badly at it?

Think about it. UFOs are supposed to be spacecraft that have crossed the vast and hostile distances between the stars. They possess powerful technologies allowing them to defy the known laws of physics. And yet, with all their abilities, they can't turn their headlights off.

If these aliens really wanted to remain hidden from us, if they really didn't want us to see them, then how come we keep seeing them . . . almost. Have they sent us their D team, the one that doesn't know which button engages the cloaking field? Are they just a bunch of Zorgovian teenagers who stole their parents' saucer and are out for a joyride?

It just doesn't make sense.

Also our cameras have come a long way since 1947. Why, then, are the vast majority of pictures of UFOs and UAPs fuzzy blobs or unfocused blurs? This point was sharpened in February 2023 when the US government shot down a Chinese spy balloon. A few weeks later, an image taken from inside the cockpit of a U-2 high-altitude jet was released. It's basically a selfie, and the image is super clear and sharp. Along with the pilot's head, you can also see the spherical balloon and make out details on the payload below, including its solar panels deployed. After seventy years of UFO sightings, shouldn't there be stacks of these kinds of pictures? Instead, the prevalence of blob and blur images serves as evidence that what's being seen are artifacts of the environment or the cameras themselves and not real physical things that should leave clear images on film or in electronic memory.

That's where we will leave it in our thinking about UFOs and UAPs. I want to make it clear that I am completely in favor of a full, open, and transparent scientific study of the phenomena. It is good that the stigma surrounding sightings by pilots has dropped off. This allows us to get a better handle on how often weird stuff is being seen in the sky. From a scientific perspective, that's a first step to understanding what the weird stuff is. While I mostly see the issue being one associated with national defense, as a scientist I am committed to remain agnostic until the data are definitive. To do science, we need real scientific data. That's the stage we may be just entering now, and we'll see where it takes us.

But as a scientist deeply interested in life beyond Earth, I don't think Earth is the reasonable focus of the question. If you wanted to find Nebraskans, would you go searching in some remote village in the Himalayas? Probably not. Instead, you'd go to Nebraska, where the Nebraskans live. Likewise, if we are looking for alien life, we need to be looking at alien planets. That was not possible before, but it is now. We're finally ready and finally capable of looking at distant worlds for signs of life. Where and how we're going to do this, and what we might expect to find—that's what the rest of this book is for.

CHAPTER 5

Cosmic Curb Appeal?

WHERE TO LOOK FOR ALIENS

When Frank Drake carried out his groundbreaking seti search in 1960, he was in the dark about cosmic real estate. Where was the best place to look for signs of alien civilizations? The most intelligent guess he could make was to pick nearby stars that looked like the Sun. But life doesn't form on stars; it forms on planets (the surfaces of even the coolest stars are so hot they'd rip life down to its atoms). In 1960 nobody knew if any other planets existed outside our solar system. In fact, nobody knew that in 1970, 1980, and 1990. By 2000, however, things had changed dramatically. Humanity had its first extrasolar planets, or *exoplanets*, in the bag. Suddenly the universe looked a lot more hospitable for life. We'd found where alien life might start and where it might thrive, and that changed everything. We were finally ready to go (alien) house shopping.

Over the last couple of decades, we've gone from knowing about just a handful of exoplanets to building an extensive census of alien worlds. The results of these studies have been eye-popping. We've found worlds that might be made of diamond, worlds that might be all ocean, worlds that could be covered in ice. Thinking about these planets as actual places, existing right now as you read these words, where winds blow through canyons and snow drifts across valleys and waves wash up on shorelines is enough to blow open the horizons of your mind. The added possibility that life in all its creativity might have taken hold on some of them can give you a case of cosmic vertigo.

The solar system also has its own opportunities for harboring life beyond Earth. Even though Mars is a frozen desert with little atmosphere now, it was once a world where liquid water rushed along the surface and gathered into vast lakes or even shallow oceans. It's possible life formed on Mars before that world lost most of its atmosphere four billion years ago. Because Mars has been part of the search-for-life game for so long, we won't cover it in the chapters that follow. (We've sent so many landers and rovers there that the planet is sort of robot-inhabited right now.) As we'll see, though, there are other places in the solar system, like the ocean moons of Jupiter and Saturn, that might harbor life, and they may tell us something about the ocean exoplanets, which may be common.

Because it's life we are most interested in, we're going to get started by making the connection between life and planets. The first stop on our tour of cosmic real estate is the question of *abiogenesis*. How does life get started on a lifeless world?

THE ORIGIN OF LIFE
The Miller-Urey experiment and abiogenesis

Life is weird.

I don't mean your particular life and whatever particular weirdness it's been manifesting lately (I'm sure it will all work out). Instead, what I mean is that life as a phenomenon in the universe is weird. Sure, black holes and their ability to bend space and time are strange. And yes, quantum physics is full of head-spinning paradoxes. But neither of these can hold a candle to the insanity going on in every cell of your body *right now*. For example, at this moment there are millions of little nanomachine tow trucks pulling molecular cargo along self-assembling railroad lines that cross the expanses of your invisibly tiny cells. That's just one example. I could go on for a long time.

There simply is no other system, no other kind of thing in the universe that can compare to life. It's life's ability to create and innovate that's unlike anything else. But how did it start? If life is so strange and different, how did it ever begin? If we want to know about life elsewhere, we need to know how it starts anywhere. That means understanding something about how life started here on Earth. Of course, life on other planets could be different. It could use a different chemical basis or different evolutionary principles. We'll consider that possibility eventually. But since we have only one example of life, it makes sense to start with Earth.

The study of life's origins is the study of *abiogenesis*—the creation of life from non-life. The goal of abiogenesis research is to understand how the basic laws of physics and chemistry allowed a bunch of dead chemicals to somehow combine and make living organisms. It's a tall order, but over the last seventy years, remarkable progress has been made in laying out the basic requirements needed to solve the problem. To get a handle on abiogenesis, let's start by thinking about what we mean when we say the word *life*.

One hallmark of life is *reproduction*. Life makes copies of itself. Life also always involves *metabolism*. It consumes energy to keep itself going. Life is also *sensitive* to its surroundings. Finally, life has the ability to change. It can *mutate* from one cycle of reproduction to the next, which is the basis of its ability to evolve. And all of life's activity is built from molecular activity. Life's weirdness begins at the molecular level.

Not just any old molecule will do, however. Life on Earth draws only from a restricted set of molecular Lego pieces to do its work.

Specifically, organic chemistry mostly involves combinations of carbon, hydrogen, nitrogen, oxygen, phosphorus, and sulfur. This doesn't mean that life elsewhere couldn't use a different molecular scaffolding, but it does mean that some kind of scaffolding using a restricted set of atoms is likely.

So, we have reproduction, metabolism, sensitivity, and mutation as the list of what life does. How does it get these done at the molecular level? In what follows, I'm going to give you an overview of a few important parts of that story. It's incomplete because it's a story that is very rich, but you can use it to impress people at your next party (unless there's a biologist around, in which case, just slink away quietly).

Most of the functions of your body are carried out by proteins, which in structural terms are polymers. This is a fancy way of saying they are long chains made of smaller subunit molecules called amino acids. Proteins get built out of amino acids via insane little factories called ribosomes in your cells. These ribosomes know exactly which amino acids to combine together to make the proteins you need exactly when you need them. How do ribosomes know what to connect? The information required to guide all this protein building is stored in another kind of molecule called deoxyribonucleic acid (DNA). The details of how the information in DNA gets read off and carried to the ribosomes' factories are head-spinning in their complexity. Life at the molecular level is remarkable and kind of crazy. That's what makes the science behind it so shocking and awe-inspiring. How the hell did all of this happen? How did a bunch of tiny molecules bouncing around at random end up acting like a computer code that stores information and self-replicates?

Back in the 1920s, Russian scientist Alexander Oparin and British researcher J.B.S. Haldane invented the modern theory of abiogenesis when they proposed that life began with precursors of biological molecules that formed through natural processes occurring on the young Earth. The idea was simple: if you can just get all the basic elements of biochemistry in the same place and let them rattle around

COSMIC CURB APPEAL?

long enough, they should combine into a form that begins to self-replicate. Although the idea is simple, it was very controversial at the time. No one had any proof that this kind of automatic assembly of life's precursors could occur through natural processes. Of course, no one had any scientifically testable idea of how life started back then, so it was all new ground. Then in the early 1950s (seems like all the important stuff for this book started in the 1950s), a stunning experiment showed that Oparin and Haldane had been on the right track.

In 1953, two chemists at the University of Chicago, Stanley Miller and Harold Urey, created a simulated version of the early Earth in a test tube. They took simple molecules like methane, ammonia, hydrogen, and water and stuck them in a jar. This mixture was supposed to mimic Earth's early atmosphere. Then they added energy in the form of an electrical current. This was their laboratory version of lightning, something that could be expected in any atmosphere. Finally, they let the results collect at the bottom of a beaker, their version of a pond of water lying somewhere on the young planet.

After running this experiment for a week, Miller and Urey found that a "brown goo" had collected in their mock pond. Chemical analysis showed that the goo was a rich soup of prebiotic molecules like glycine, lactic acid, and urea. All of them were important molecules that life uses quite a lot of. Most important of all, though, a substantial fraction of the molecules floating in the brown goo were amino acids. The basic stuff of life's proteins were just sitting there in the bottom of the flask. It almost seemed that all the scientists had to do was just wait a bit longer and an amoeba would have crawled out of the beaker. The ancient problem of how life started seemed a step or two away from being solved!

The Miller-Urey experiment was a major breakthrough, a major surprise, and an instant classic of biochemistry. For lots of scientists, it was proof that you could start with nonliving material and the natural processes of physics and chemistry would eventually serve up the molecular basis of life. Many were ready to claim victory.

It turned out to be a little more complicated. OK, more than a little. Later studies made it clear that the early Earth would not have had an atmosphere like the one that Miller and Urey had assumed. Later, scientists realized that other environments might serve as better starting points for life than warm ponds with a primordial soup of prebiotic molecules. Some researchers, for example, focused on volcanic vents deep in the ocean, where molten rock comes in contact with sea water, creating a stew of chemicals just right for getting biology going. The hot water is important too, because heated molecules collide with each other faster and more often than chilly ones. The more random collisions, the faster the whole process works.

Just as important, though, researchers soon realized that proteins weren't enough. The real problem was getting replication going. Calculations showed that assembling DNA from random collisions can take longer than the age of the universe. Not good. That's why scientists began thinking that the first molecule to copy itself need not be DNA. Their focus then turned to RNA (ribonucleic acid). RNA is like one strand of the DNA double helix with a slight chemical difference in the bases. RNA is central to the machinery of life because it copies the instructions from DNA and transfers the information to the protein-factory ribosomes. By turning to RNA, biologists began having more success. These "RNA world" models begin with computer simulations of environments containing lots of tiny RNA strands that can be assembled by collisions in less than a second. The models then show that the same collisions can construct much longer and, more important, self-replicating RNA sequences in about three hundred thousand years. That may seem long to us, but it's a blink of an eye in Earth's early history.

From the Miller-Urey experiments to modern RNA-world computer simulations, success in abiogenesis studies may be more a matter of when than if. It's clear that nature knows how to create the building blocks for life. Amino acids have even been found floating around in interstellar clouds! The question then becomes how exactly these building blocks get combined through random collisions into something that can reproduce itself. Once replication begins, the game of life can really get started. By creating so many copies of itself, the replicator will take over and soon outcompete all the other chemical processes trying to build up concentrations of other chemicals. Now throw mutation and evolution into the mix, and the replicator can begin making better and more interesting versions of itself. Scientists have even seen how lipids can form little bags or protomembranes as they float around in pools of chemicals. Perhaps some early version of a replicating molecule started using these lipid membranes to protect itself four billion or so years ago, creating an early protoversion of a cell.

The key ingredient in this whole story is time. Human scientists can't wait more than a few decades to see whether their experiments pan out. Nature is not so impatient. The prebiotic soup could percolate for hundreds of thousands of years, with countless molecular combinations forming, until self-replication randomly occurred. When people question whether life could have formed from non-life, they often fail to take into account the vast stretches of time that were available for the process to work. We are talking hundreds of millions of years here. That's a million times longer than you will be alive.

Thanks to the Miller-Urey experiment and the work that has followed, there's a solid scientific foundation for understanding how life can emerge from non-life. There are significant mysteries still to be understood, but the basic possibility of abiogenesis is no longer mysterious. It gives us firm ground to stand on as we look back and understand how life formed on Earth or look out and understand how it might or will have formed on alien worlds. With this molecular perspective, we should be in a strong position to understand the engines of alien life when we find evidence for it.

· · ·

THE OCEAN MOONS
Who knew?

If the hunt for life is also the hunt for water, there's a whole other class of target hiding out at the edges of the solar system. Consider for a moment all of Earth's oceans—the vast depths of the Atlantic from Newfoundland to Ireland in the north and the tip of Argentina to the Horn of Africa in the south. And don't forget the endless expanse of the Pacific with its deep canyons stretching down six watery miles. Now hold the thought of all that water in your mind and consider this crazy solar-system fact. All of the surface water on Earth is just about half the amount in the ocean on Europa, a small moon of Jupiter. And Europa is just one of the solar system's *ocean moons*.

Jupiter and Saturn are themselves not a good place for life to form. These huge worlds have no surface. They are just endless depths of hydrogen and helium gas that reach pressures that could crush submarines (and then even diamonds) as you go deeper. But the gas giants are surrounded by extended families of orbiting moons. Most of these are nothing more than captured asteroids. A handful of the moons are, however, small worlds in their own right. In 1979 NASA sent probes flying past Jupiter, and the high resolution images they returned of Europa, the second-closest moon, were nothing less than shocking.

By comparison to most other objects in the solar system, Europa's surface was as smooth as a baby's butt. Unlike most other moons, there were almost no craters. Instead, the images showed a network of fractured lines that looked a lot like cracks in arctic ice. And that's exactly what they were. Astronomers soon realized that, unlike the other moons orbiting Jupiter, Europa was entirely covered by a layer of ice at least six miles thick! Below that mantle of ice was a liquid water ocean stretching down as far as sixty miles. Europa is a rocky moon entirely covered in a deep ocean topped off by a thick mantle of ice.

COSMIC CURB APPEAL?

The realization that Europa had so much water upended astronomers' expectations for life in the solar system. Jupiter and Saturn are far outside the Sun's habitable zone, so no one ever focused much on them as a home for life. But once the ocean on Europa was discovered, scientists had another world within the solar system that hosted life's most important resource. Suddenly the question of life on Europa loomed large. But no sunlight could ever penetrate Europa's six-mile-thick icy crust. Without light, what would power a Europan biosphere?

The answer is one of our other forces: gravity.

Jupiter is a huge planet with 317 times the mass of Earth. The gravitational pull from all that Jovian stuff tugs and stretches the insides of Jupiter's moons, including Europa's. It's the same process that creates the ocean tides on our Earth, which are mainly driven by the Moon's gravity. Tidal forces from Jupiter stretch and squeeze the rocky core of Europa as it swings around in its orbit. All this stretching and squeezing heats up the moon, melting some of its rocky interior. That heat is an energy source that might power Europan life. On Earth, life may have started in deep-sea vents where heat and molten rock boil up through the planet's crust. The same process could have happened, or could be happening right now, on Europa. Deep in the dark world of that moon's ice-encrusted oceans, there might be vast ecosystems powered by thermal vents. Who knows what kinds of creatures evolution might have created, living down there off the heat and crazy chemistry made possible when molten rock hits liquid water. As exciting as this idea may be, Europa is not the only place where this kind of drama may be playing out.

Enceladus is a small moon orbiting the ringed gas giant Saturn. In 2005 the Cassini space probe orbiting Saturn snapped images that shocked the astronomical world. Plumes of water a hundred miles high were seen exploding into space from deep cracks in Enceladus's southern hemisphere. Another subsurface ocean had been discovered. When scientists ordered Cassini to fly through the plumes, the spacecraft's instruments revealed that the water was actually a *brine*, water mixed with salt and other compounds. Adding salt to water changes it in ways that scientists think are important for the origin of life. A salty brine will, for example, resist freezing better than salt-free water. That could extend its ability to host biochemistries even in really cold environments.

So, is there life on these moons? God, I hope so. That would be so cool. But I'm not gonna lie to you; finding that life won't be easy. The only way to know for sure may be to send landers across seven hundred million miles to Jupiter and nine hundred million miles to Saturn. Then what? Do they drill their way through miles of ice? Melt their way through? It's hard to even imagine the technology that could do either. Only time, ingenuity, and funding will tell which path is best.

Here's the main thing: if these ocean moons exist in our solar system, they probably exist in others too. That opens up a lot of possibilities for life and even intelligence. Could technological civilizations emerge in these icebound ocean moons? What would they be like? How would they imagine the universe when they can't even see the stars? Do they ever manage to escape the ice? These questions live at the frontiers of science and our imaginations. What we can say for sure is that the ocean moons in our own solar system are real and they raise the stakes for life in the cosmos in ways no one could have imagined before.

EXOPLANETS
The revolution will be telescoped

"Do other planets exist outside our solar system?" was a question as old as the one about life. For 99.9 percent of human history, no one knew if our solar system with its central star and family of worlds was a freak or the norm. It seemed entirely possible that planets like Earth or Jupiter were really hard to form, making our solar system a kind of ultra-rare jewel in the galaxy. On the other hand, maybe planets were easy to create, making our solar system as common as dirt. The only way to tell was to find a planet orbiting another star.

Planets are pretty small, and stars are pretty big. That sums up why finding planets outside of our solar system was so hard. Trying to directly image a planet orbiting a distant star is basically the same as looking from San Francisco to New York City and picking out a firefly flitting around a light at Citi Field (Go, Mets!). Good luck with that. It's a problem so hard that people tried for centuries to find other ways to detect exoplanets. They failed, a lot.

Because life starts on planets, finding planets beyond our solar system has always been an astronomical priority. History is full of people ruining their astronomical careers by claiming they'd detected an exoplanet. Even as late as the 1960s, scientists were falling on the sword of exoplanet detection claims that no one else agreed with. What changed that allowed astronomers to finally "see" one? Precision technology. The first exoplanet was discovered by watching the *star* swing back and forth a trifle due to the relatively minute effect of planet's gravity. This "radial-velocity" method required insanely sensitive instruments. It also required a kind of planet that we don't have in the solar system. Once we started finding solar systems other than our own, it turned out that *we* were the weirdos.

If you recall your fifth-grade Earth Science class, the solar system has a tidy arrangement of inner "rocky" planets and outer "gas-giant" planets. The inner planets are relatively small and have tight orbits packed close together. Mercury takes just 88 days to complete an orbit while Mars takes about 687. The outer gas and ice giants are *much* bigger with widely spread-out orbits that can take many decades to complete. With nothing else to go on, it was reasonable to think that most solar systems would have the same kind of architecture. Nature, however, is not reasonable.

The first exoplanet discovered, called 51 Pegasi b, because it orbits the star 51 Pegasi, was the size of Jupiter. Its orbit took just four days to complete, making it almost ten times closer to its star than hellscape Mercury is to our Sun. Scientists creatively dubbed this new kind of planet a hot Jupiter. Being both massive and on crazily close orbits favored detection of hot Jupiters because they induce the biggest wobble in the host star. That makes them easy to see with the radial-velocity method. Soon lots of other hot Jupiters were getting discovered.

Then things got even weirder.

As the 1990s rolled into the 2000s, new planet-finding techniques were added to astronomers' tool kit. The most important addition was basically looking for the exoplanet version of an eclipse; when a planet passes in front of its sun, it will block a wee bit of the star's light. Astronomers call this mini eclipse a transit, and by the second decade of the new millennium, astronomers were using the transit method to make wholesale detections of exoplanets. The flood of new discoveries was mainly the result of a plucky little space telescope called Kepler. Before the Kepler mission, astronomers had discovered a few hundred exoplanets. Just three years after its launch, Kepler bumped that number up into the thousands.

Many of the newly discovered planets were eye-popping examples of the bizarre. Most stars in the universe are smaller and much dimmer than the Sun and are therefore called dwarf stars. Habitable-zone planets around dwarf stars are so close that they take only a few weeks to a month to complete an orbit (i.e., their year could last as short as a week). They are also so close that on the planets' dayside, the stars are huge in the planet's sky, compared to our Sun, which appears about the size of a quarter held at arm's length. Speaking of dayside and nightside, one of the weirdest things about dwarf-star habitable planets is that the sun never moves in their sky. A dwarfstar planet in the habitable zone may be so close to its star that the star's gravity locks the planet's spin to its orbit. That means the same side of the planet always points toward the star. This translates into the sun just sitting in the exact same place all the time on the dayside of the planet. On the nightside, it's always night, forever. Any life that forms on these planets, (if life *can* form on them) would evolve with no day-night cycle. Can there be life at all on the dark side, where there is never ever any sunlight? What about climate? How does the weather work on a world where it's always noon on the dayside and midnight in the other hemisphere? How about photosynthesis? Could plants evolve to harvest starlight on planets orbiting a dwarf star? Such stars are a lot redder than our yellow Sun, that color difference being why we call them red dwarfs. Red-dwarf planets showed astronomers just how extreme even habitable-zone exoplanet environments can get.

Most important of all, though, was the emerging planetary census. By the early 2010s, astronomers had enough data to start cranking out the statistics of planets. Whereas no one knew if there were any planets orbiting any other stars in 1992, by 2012 humanity was certain that pretty much every star in the sky hosted a family of worlds.

Every. Single. Star.

Think about that the next time you look at the night sky. Almost every star you see has at least one world orbiting it. Most probably host full families of worlds. Even more amazing, thanks to the new planetary demographics, astronomers were able to say with certainty that about one in five stars hosts a planet in a "habitable" orbit, i.e., an orbit where liquid water could exist on the surface and therefore where life could form. When you go out at night and look at five stars, one of them likely has a rocky world where fog could be drifting across lowland valleys, waves could be washing up on shorelines, and maybe, just maybe, someone like you could be soaking up starlight.

The exoplanet revolution changed everything in our search for alien life. It vastly increased the probability that such life exists and told us exactly where and how to look for it. Finding exoplanets is one of the greatest achievements of human culture and has put us firmly on the path to what will be, if it happens, the greatest discovery ever made.

PLANETS GONE WILD
The super-Earth enigma

It's hard to know what to expect when you have nothing to go on. For the whole of human history, all we've ever seen of planets are the examples in our solar system. Based on that small experience, how can we say what's common and what's unusual? Is the Earth an average-size planet? Is its location from the Sun the kind of thing we'd expect from most planets, particularly the ones where we expect life to form? It's easy to assume that what you know is normal until you know more.

As each new exoplanet got added to the growing census, astronomers could better see which kinds of planets were common and which kinds were rare. After a few years, they had enough data to define the galaxy's average planet. What they found was very surprising. There *is* a kind of planet that's pretty common, but it's also a kind of world we don't have in our solar system.

To get a good picture of the galaxy's average planet, let's remember what our own little corner of the universe is stocked with. Our solar system has eight planets. (Please don't start with me about Pluto, OK?) Those eight planets come in three basic classes. First, there are the "terrestrial" worlds: Mercury, Venus, Earth, and Mars. These are rocky planets with relatively thin atmospheres (meaning that the atmosphere doesn't weigh much compared to the rest of the planet). The most massive terrestrial world is Earth. (Go, Earth! We rule!) But the real heavyweights of our solar system are the gas giants, Jupiter and Saturn. These are, as the name implies, giant balls of hydrogen and helium gas. Jupiter is over three hundred times more massive than the Earth and eleven times larger in diameter. Saturn comes in at about a hundred Earth masses and ten Earth diameters. Finally, we also have the ice giants. There's Uranus (stop laughing) and Neptune. These worlds have an upper slush of water, ammonia, and methane ices with a pretty big rocky core at the center. Uranus and Neptune are nearly equal in size, Uranus being about fourteen times more massive than Earth and about four times wider.

The important takeaway from this little tour is that there are no planets in our solar system between one and fourteen Earth masses—nothing, nada, zilch. If you didn't know better and had only our solar system to go on, you'd think that nature doesn't *make* planets in that size range. If you thought that, you'd be spectacularly wrong. Do you want to guess the mass of the universe's most common type of planet? Yup, it's right in that gap between one and fourteen Earth masses. The cosmos has a sense of humor, apparently.

Astronomers can now say with a lot of confidence that most planets have masses somewhere between one and ten times that of the Earth. Why does this matter? It matters for life. Keep in mind that giant planets (gas or ice) and rocky planets are really different. Astronomers are pretty sure life's never going to form on a giant planet. There's a laundry list of reasons for this conclusion, chief among them being the fact that gas and ice giants don't even have a surface. The pressures and temperatures on these giant worlds get so high that, while we can't rule out the possibility of life forming on them, it seems like a much longer shot compared to planets that have surfaces with liquid water on them.

Astronomers now have this entirely other kind of planet to figure out. They call these worlds *super-Earths* because they have masses larger than Earth but less than Neptune. What exactly *is* a super-Earth, and most important, is it a good place for life? This last question is the really important one. If super-Earths are the most common kind of planet, it would be really great if they were cozy places to get biology going.

The exciting answer to both questions is that nobody knows. Super-Earths could be places with rocky surfaces and relatively light atmospheres. In this case, they would be scaled-up versions of our world. If life evolved on this kind of super-Earth, it would have to contend with the planet's high gravity. This might mean that every animal would have thick legs like an elephant or be built low to the ground like a crocodile. That would be pretty cool. But because the atmosphere would get pulled tighter to the surface, the air on a super-Earth might be so thick that you could fly by just waving your arms (kind of). That would be pretty cool too.

However, super-Earths could also tilt more toward the ice giants, with massive, thick outer layers of mixed gas and ice. The cores of this kind of world might not be molten rock and iron like Earth but might be made of water under such insane pressures that new kinds of ice form. This kind of super-Earth does not, at face value, seem like a good place for life to get started.

The problem right now is that astronomers can't yet tell which kind of planet most super-Earths will turn out to be. We've only just started observing them with next-generation telescopes that can see deeply enough with enough resolution to help us figure out what category of planet super-Earths belong in. We also can use banks of super-high-powered lasers to do laboratory experiments re-creating conditions inside super-Earths (this is something my colleagues and I do at the University of Rochester's massive Laboratory for Laser Energetics). But the fact that we don't have a super-Earth in our solar system to study up close and personal makes the problem very challenging. Still, there is nothing a scientist loves more than a good challenge. Will super-Earths be super-friendly places for life to form? Being the most common kind of planet in the universe, will they be the kind of planet we might expect to host alien civilizations? If so, how will the super size of these super worlds affect the kind of life and civilizations they nurture? These are the questions waiting for us as these remarkable and unexpected planets come into focus.

• • •

SNOWBALL WORLDS AND OCEAN WORLDS
Winter is coming, and so is the flood

Sometimes, when I want to play hooky, I head over to the Rochester Museum and Science Center to look at my good friend, the giant hairy elephant. Being a mastodon, he's been dead a long time, but that doesn't make him any less impressive. Ten feet tall, six tons huge, thick coat of matted fur, giant tusks. He and his kin used to be legion in these parts. Now his kind are gone forever. I keep visiting him to remind me of something that really gives me the willies—ice ages. Just twenty thousand years ago, my little upstate New York city was sitting at the bottom of a glacier that stretched two miles into the air. Let me say that again, so it really sinks in. Rochester, New York, used to be under almost two full miles of solid ice.

Damn!

Ice ages are planetary events. Over the last few million years, Earth's northern regions have spent most of the time almost entirely covered by glaciers. These ice ages can last for a few hundred thousand years before the glaciers retreat (melt). In the ten-thousand-year interglacial periods between the ice ages, interesting stuff can happen—like, say, the whole of human civilization. Glaciate, melt, repeat. That's an ice age.

The idea that a miles-thick slab of ice could extend from Seattle all the way to Long Island for a hundred thousand years is pretty freaky. The idea was so freaky, in fact, that it took geologists a while before they accepted the evidence that ice ages were a thing in Earth's past. Now, however, Earth scientists are facing another come-to-Frosty moment. Long before the current string of ice ages, there have been times when the entire planet was locked in ice. These periods, called Snowball Earth phases, are a radical example of climate change, and they may have had profound effects on our planet's life. Evolution faces serious challenges when the whole globe is frozen over. As important as they are for the story of life on Earth, these maximum icebox phases matter in our search for alien life too. If it happened here, it could happen anywhere, and that means astronomers must now ask if the galaxy is awash in snowball worlds.

There's evidence that Earth has experienced three distinct snowball phases. Two were "relatively" recent, occurring between three-quarters and one-half billion years ago. The longer of these snowballs may have kept Earth locked in ice for about a hundred million years. Long before that, there may also have been a snowball phase about two billion years ago lasting even longer. During these very long winters that put *Game of Thrones* to shame, all continents would be buried miles deep in glaciers. The seas would fare no better, getting topped by a layer of solid ice at least thirty feet thick.

The onset of a snowball phase is a dangerous moment for a planet. If ice manages to cover an entire world, it can become difficult to make it ever melt again. Understanding the physics, chemistry, and geology of how this works comes from the field of planetary climate science. Studies of Mars, Venus, and Earth have given researchers remarkably deep knowledge of how any climate on any planet will work—and also an understanding of the climate change we humans are driving now. In practice, this means understanding how sunlight, a planet's rotation, its atmospheric and ocean currents, and its chemistry all come together to set the world's "climate state." These states are the average weather conditions across a planet for thousands to millions of years. For example, human civilization rose up in the geological epoch called the Holocene, which has lasted for about ten thousand years now. The Holocene's climate state has been relatively warm and relatively moist, which is great for things like stable agriculture. During ice ages the climate state is different. The Earth becomes mainly cold and mainly dry. The dry part comes because lots of the water is locked up in glaciers. During the ice ages, so much water gets frozen into glaciers that sea levels can drop four hundred feet below where they are today. That's the height of a forty-story high-rise.

COSMIC CURB APPEAL?

Once a snowball phase gets going it can, well, snowball. The increasing snow makes the planet more reflective. Sunlight bounces back into space rather than warming the ground, and the planet temperature drops. In other words, more snow falling creates the conditions for more snow to fall. In climate science, this is called a *runaway effect*, and it explains how a planet can eventually get trapped in the frozen state. The only way to warm the planet back up is to get more greenhouse gases into the atmosphere via volcanoes and other processes. That seems to be what saved the Earth. Volcanoes belched enough carbon dioxide into the air to slowly bring the temperature up and melt the glaciers. But climate models of exoplanets show that Earth may have gotten lucky. Computer simulations of alien Earthlike planets show that, once a planet drops into the icebox, it can stay there forever. If this is true, then snowball worlds may be common even for entirely Earth-like planets in their habitable zones. In the movie *Interstellar*, the planet where Matt Damon's evil character gets trapped seemed to be a snowball world.

Can there be life on a snowball world? If it started before the snowball state arrived, could it still survive and flourish? Better yet, could life evolve toward a technological civilization on a snowball world? It seems possible that microbial life might not have a problem with living in an icebox. Scientists find lots of critters living in the Antarctic. Complex animals may be another story, though. If a planet goes snowball before complex, multicellular life gets started, the cold may become a kind of energy barrier preventing larger life-forms from appearing. Evolution might favor smaller, less heat-hungry organisms. This would mean that worlds trapped in a snowball might not develop civilization-building aliens. If, however, the planet went snowball after an intelligent species emerged, it might be possible for them to find ways to continue to live and even thrive. There is no inherent reason why such abominable snow aliens couldn't develop technology if they access the necessary resources. Perhaps our first contact will then be with aliens whose spaceships resemble traveling refrigerators.

The amount of ice a planet hosts depends, of course, on how much water there is. A desert planet could never become a snowball world. However, as the exoplanet revolution unfolded and astronomers built up a census of the thousands of new worlds they'd discovered in the galaxy, they realized that one feature of our Earth looks really, really weird.

It's blue, *and* it's brown.

About three-quarters of the Earth's surface is ocean. The rest is land. That means, once you fill up all the planet's low-lying basins with the Earth's inventory of water, the continents still rise high enough to avoid getting flooded. Other than rainfall, rivers, and some lakes, the dry land stays dry. That's where a whole lot of critical steps in evolution played out, including the development of big brains for civilization building.

Why did Earth get only that much water and no more? If Earth's store of H_2O had been higher, then—other than a few mountaintops—there would be no dry land at all. Ours would have been an ocean world. Astronomers now think this kind of planet may be far more common than the mix of land and water we ended up with.

To understand why ocean worlds can exist, you must ask how worlds get their water in the first place. Planets are born from solar system–size disks of gas and dust that surround newly formed stars. Terrestrial planets (i.e., rocky worlds like Earth) get built slowly as dust particles in these disks collide to form orbiting rocks. The rocks then collide and form boulders, which then collide to form asteroid-size bodies (tens of miles across) and so on, all the way up to full-scale planets like Mercury, Mars, Earth, and Venus. The whole process takes about ten million to one hundred million years, and a lot of heat gets generated in all these collisions. Whatever H_2O molecules might have been around at the beginning of the planet-building process tend to vaporize away once the planet has fully formed, so most rocky worlds are probably born bone-dry. Whatever water they are destined to end up with as a mature world

COSMIC CURB APPEAL?

must get *delivered* after they form. The delivery agents are comets and asteroids.

Comets, in particular, are the universe's slush balls. A typical comet can hold many billions of tons of ice with some rocky material mixed in.[1] It would take millions of comets hitting the Earth to deliver enough water to fill up the oceans. If that seems like a lot to us, that's only because our imaginations stumble at the incredible violence of Earth's early history and the vast spans of time involved. Asteroids can also contain a lot of water. During its first half billion years, the Earth was continually pelted with solar-system construction debris, like comets and asteroids. If almost any one of these collisions happened today, it would wipe us off the face of the planet. Back then, though, before life appeared, periodic comet and asteroid impacts were the norm. Even a few comets or asteroids delivering their store of water every thousand years or so would have been enough to create the modern oceans.

Why stop at just a few comets every thousand years? It's not hard to imagine another solar system where twice that amount gets dropped onto a young rocky planet. Or why not five or ten times more? Histories like this would lead to so much water dumped onto a planet that it must end up as an ocean world. Any and all continents would get drowned. On the other hand, reduce the number of comets hitting the planet by a factor of ten from what Earth experienced, and you'd never get anything more than some lakes. In that case, you'd end up with a desert world, like Arrakis in the novel *Dune*. Put all this together, and you're left with the conclusion that there are lots of ways to make desert planets and lots of ways to make water worlds. But something like Earth, however, requires a Goldilocks history with a "just right" amount of water delivered by comets or asteroids. Of course, any mention of Goldilocks in astrobiology always leads us straight to the most important question: What about life and civilizations?

Based on everything we understand about planets and their

formation, we can say with confidence that ocean worlds must exist, and they must be stunning. An ocean world would be an endless expanse of water that might reach depths of hundreds of miles or more. (Earth's oceans never get deeper than about six miles.)* Astronomers and planetary scientists are just beginning to explore the possibilities that come with such deep continuous oceans. Do these worlds host planetwide storms? Are there hurricanes that engulf half the world? What kinds of vast currents might flow in such deep ocean? Most important of all, can life form on an ocean world? And if it does appear, can technologically sophisticated species emerge without land? We've already encountered these questions with ocean moons, but those are small satellite worlds that may not have atmospheres. Ocean worlds can have lovely blue skies and night-times full of stars for a young civilization to see.

Astrobiologists have strong reasons to think that water is a prerequisite for making life anywhere. Desert planets are a long shot for getting biology going and, more important, for getting a rich enough biosphere to lead to intelligent, tool-building species. Ocean worlds will certainly have enough water to go around, but without any continents, can they get life going *and* push it along toward a high-tech civilization?

Getting life started may not be too much of a problem. Earth's first replicating biological systems may have formed deep in the ocean at geothermal vents. That's where hot magma from the planet's interior breaks through the seafloor, providing a source of energy and chemicals to get life started. For much of Earth's history, evolution was a story playing out in the oceans. The "settlement" of the land by life didn't really get going until billions of years after life appeared.

Using Earth as a model, it seems that even a planet entirely covered by water could get biological evolution started and take simple organisms to ever more complex forms, including some with high intelligence. On Earth the octopus, that most alien-looking of

---

* Just a reminder that ocean *moons* will be satellites of other planets, like Jupiter-size gas giants. An ocean world would be a full-fledged planet orbiting a star on its own.

species, has developed impressive cognitive capacities, including the ability to make and use tools. Just as impressive, the octopus accomplishes these feats with a brain that looks nothing like our own. Each tentacle has its own version of a brain, which must communicate its intentions and decisions to the central brain. This makes the octopus a kind of multiple-personality creature. If this kind of intelligence could evolve in the oceans of Earth, it certainly seems a reasonable bet that ocean worlds might have their own intelligent species. Some of these may develop rich, complex cultures that build cool stuff, like vast hanging cities using kelp-like or coral-like materials. The possibilities are as endless as they are awesome.

There is, however, a big fly (or sardine) in the ointment when it comes to really advanced civilizations on ocean worlds. If the oceans on a planet are deep enough, then pressures at the sea floor may be so high that water freezes into entirely new and weird forms. An ocean bottom covered in a ten-mile-thick layer of ice-9 (or whatever these new ice forms will be called) would mean no hydrothermal vents and no molten rock with liquid water for launching biology. Moreover, even on ocean worlds where life did form and go on to evolve intelligent species, a world of water is a world without fire. A world without fire is a world without metallurgy. If an intelligent species has no easy way to melt and combine metals, would it ever progress to making computers or radio telescopes or rocket ships? Without fire, could there be industry of any kind?

These questions don't have answers yet, but we came to recognize the existence of ocean worlds and their profusion only within the last twenty years. Over the next fifty, we will have more data and more insights. Perhaps there are many ways for technological civilizations to bypass the need for fire on ocean worlds, allowing them to eventually build high technology and even leave their planets. If so, given how common ocean worlds may be, perhaps the first space travelers we encounter will live in ships filled with salt water rather than oxygenated gas.

TEN BILLION TRILLION CHANCES TO ROLL THE DICE
The pessimism line and what it tells us

For all of human history, we didn't know if there were any planets orbiting other stars, but now we know they're everywhere. What, exactly, does that knowledge buy us? Have we really gotten closer to what we want? If the question we're really interested in is about *aliens*, not alien planets, then what does the exoplanet revolution change when it comes to actual aliens?

Funny you should ask, because this is exactly the question my former teacher Woody Sullivan and I went after, back in 2016, as the exoplanet census started to solidify. Woody is a radio astronomer at the University of Washington and was part of the second generation of SETI researchers after Frank Drake. Back in the late 1980s, when Public Enemy ruled the airwaves and the original *Predator* movie had just come out, I was a lowly graduate student at UW. Even though I wasn't doing SETI yet, my lifelong interest in aliens drew me to Woody's boundless enthusiasm. We stayed friends through the decades. When I started asking myself what all those exoplanet discoveries were good for relative to *exocivilizations*, I called Woody. We decided that answering the question would mean going back to—where else?—the Drake equation.

To understand what Woody and I did in our research project, let's write out the Drake equation again and remind ourselves what each of its terms means (you can also read chapter 1 again). Here it is in all its glory.

$$N = R_* \cdot f_p \cdot n_e \cdot f_l \cdot f_i \cdot f_c \cdot L$$

Quickly this time, let's go through in words what the equation says. The number of aliens we can detect (N) equals the number of stars forming each year (R_*) times the fraction of those stars with planets (f_p)

times the number of planets with an environment in which life might form (ne) times the fraction of planets where life actually does form (f_l) times the fraction of those planets that evolve intelligence (f_i) times the fraction of those intelligences that go on to create technological civilizations (f_c) times the average lifespan of those civilizations (L).

When Drake wrote this equation, back in 1961, the only term astronomers had data for was the number of stars forming in the galaxy per year (R_*). The other six terms were anyone's guess. That remarkable state of confusion lasted for another three decades, until the exoplanet revolution nailed the next two terms: the fraction of stars with planets (f_p) and the number of planets with an environment for life (n_e). It was the exoplanet census I keep yabbering on about that allowed astronomers to know *with certainty* that almost every star in the sky has at least one planet in orbit (translated into math as $f_p = 1$). It also allowed them to see that about one in five stars has an Earthlike planet in its habitable zone (in equation form that's $n_e = 0.2$).

Going from knowing just one Drake term to knowing three of them constitutes one hell of a quantum leap in understanding. That leap also meant that all the astronomical terms were done. We'd determined their actual values by making observations with telescopes. The remaining unknowns in the Drake equation all have to do with life in one way or another. All that was left to solve were the biology unknowns. They were either about the basic biology of getting life started (f_l), evolution leading to intelligence (f_i), evolution leading to civilizations (f_c), and finally, the sociology of how long civilizations last (L). We saw that our next step was to rearrange the Drake equation into a new form so we could ask a different question, one we could get an answer to, using the actual exoplanet data.

The original form of Drake's equation asks a single, specific scientific question: What is the number of technologically advanced civilizations in the galaxy that can emit radio signals detectable on Earth right now? We still didn't have the information we needed to answer that whopper, but by making a few changes, we came up with a question we *could* answer: What does the probability of forming a technological civilization have to be so that we're the only civilization that ever existed? That may sound convoluted, but the whole point of science is to answer the questions you can with the data you have.

So here's what our question really meant.

Every habitable-zone planet is like an experiment the universe runs in life and civilizations. A technological civilization either will or will not form on a habitable-zone planet, depending on a boatload of biological, evolutionary, and sociological factors we know nothing about right now. If we did know about them, then we'd also know the full probability that life and a civilization could form on some randomly chosen habitable-zone planet. We could, in other words, know the probability that any of those experiments would succeed. Let's call this the biotechnical probability, which we can represent as f_{bt}. If this biotechnical probability is close to 1, then pretty much every habitable-zone planet will form life that goes on to make a technological civilization. If f_{bt} turned out to be 0, then no planet would ever form a high-tech civilization. Of course, we exist, so f_{bt} can't be *exactly* zero.

As I said, the bummer at this point in history is that we don't know all the biology or evolution needed to get the *actual* biotechnical probability nature has set for itself. We can't, in other words, figure out fbt on our own from "first principles" in biology, evolution theory, and sociology. But the cool thing Woody and I saw was that, using the exoplanet data and our new form of the Drake equation, we could put what scientists call a limit on the biotechnical probability. We could use the data to find out what fbt would have to be for us to be the only technological civilization in the entire history of the universe. Why does that limit matter? If the actual value nature sets for f_{bt} is larger than the limit the data gave us, then we're *not* the only such civilization. The limit tells us about how likely it is that we're the only civilization ever.

COSMIC CURB APPEAL?

After gathering the data and thinking about how to rework Drake's simple equation, we found that a conservative estimate of our limit on the biotechnical probability was one in ten billion trillion. If that sounds like a really small number, it is. In scientific notation, it's 10^{-22}. This expression translates into a zero followed by a decimal point followed by 21 zeros followed by 1. You want to see that written out?

$$f_{bt} = 0.0000000000000000000001.$$

What does this number tell us? It says that the only way we're the only civilization in cosmic history is if the odds of forming a civilization on a random planet are crazy small. It means that the actual odds—the ones that nature sets with its laws of biology, evolution, and sociology—can be pretty small, and yet there would still have been lots and lots of civilizations in cosmic history. Let's say the actual odds of getting a technological civilization on a planet were one in a million. That, itself, seems pretty small for getting a civilization. But using what Woody and I found, it would still leave the universe with trillions of high-level civilizations! That's a lot of Klingons, Wookiees, and Predators to populate the cosmos with.

What Woody and I worked out was that nature has run some ten billion trillion experiments with planets, life, and civilizations over the course of cosmic history. The only way we're the only civilization in all that cosmic history would be if *every other* experiment failed. If that's the case, nature must be really biased against making civilizations. The bias would be so strong that it would fall to the pessimists to explain why it could only happen here. That's why we called our limit the pessimism line. If nature chose a biotechnical probability below the pessimism line, then, yes, we are the only civilization in cosmic history. But if nature's actual value is above that line, then we are not the first or the only.

When I first worked out that ten billion trillion number, I kinda had a "moment." I felt as if I knew something no one else did, as in, "Holy crap, there *must* have been aliens." Is that right? Well, if we're gonna be sober scientists and not fall in love with our own results (the essence of good science), we have to stop and really consider the situation. We have to admit that while one in ten billion trillion seems like small odds, the actual odds might be smaller. When we say a number is small, what are we comparing it to? Remember, we don't have anything based in a theory of biology, evolution, or sociology that tells us what to expect. Also, that biotechnological limit Woody and I found only tells us about the *history* of the universe. It doesn't tell us if any of those civilizations are around now. Maybe there have been a hundred billion Klingon and Wookiee civilizations, but they're all gone now. The universe is over thirteen billion years old. That's a lot of time for civilizations to come and go.

Still, even with these caveats (and science is all about caveats), it's hard not to have your mind blown by the numbers. Just a few decades ago, no one knew if even one stinking exoplanet was out there in orbit around a single other stinking star. Now we can say with confidence that there are roughly ten billion trillion planets in the right place for life to form. The sky is rich in worlds. The galaxy is rich in worlds. The universe is rich in worlds. Ten billion trillion habitable-zone planets is a lot of experiments and a lot of possibilities. A universe of worlds means there are probably many where snow is falling in silent canyons, winds are blowing through mountain passes, and waves are breaking on golden beaches right now. (I know I keep coming back to this point, but it's really worthwhile to spend some minutes actually imagining what's likely happening on these planets at this time.) Epicurus was correct. With that realization, we are ready to take the next step and ask if life has made its appearance on any of these worlds. And if it has, how are we going to find that life now?

CHAPTER 6

The Cosmic Stakeout

HOW WE'RE GOING TO SPY ON ET

The exoplanet revolution taught us exactly where to look for aliens. We know which stars have planets. We know which of those planets are in the habitable zone where life can form. We'll soon know which of those habitable-zone planets are Earth-like enough to be worth pointing our telescopes at. Now that we know where to look, the next question is *how* do we look? What kind of science tricks do we have up our sleeves to "see" alien life across the trillions upon trillions of miles to the stars?

This is the part of the story that really fries my potatoes. That little kid in me who dreamed of finding alien life never imagined how powerful our telescopes would become and how that power would give us the capacities to find life. Every week at the University of Rochester, we bring in other scientists to tell us about new research and blow our minds. In one talk after another, I've learned about the ways we'll be able to check for alien life on distant worlds. After every one of these presentations, I feel I need to pinch myself. Really? Are we really finally ready to do this? The answer is yes, we are really finally ready to do this. We are going to do this. It's not going to be easy, though, and there is still a lot of scientific creativity required.

Still, over the next ten, twenty, or thirty years, we're going to be spooling in data from the stars that could finally answer our alien question. I can't tell you what that answer will be, but I can tell you what we're going to be looking for. I can tell you that because it is exactly what I, my colleagues, and the entire research community that we are just one small part of are working feverishly on.

So how are we going to do it? How exactly are we going to find aliens, including the "dumb" kind (microbes) and the "smart" kind (civilizations)?* The answer: we're going to be spying. Through a series of truly stunning technological and theoretical developments, we can now find aliens just by watching their planets. We don't need them to send us messages. We don't need them to announce their presence to the cosmos. (That actually might be a really bad idea.) Like detectives on a stakeout, we can just hang out with our donuts and cold coffee, watching and waiting. The remarkable telescopes and light analyzers we've built, along with the deep understanding of how planets and life evolve together, give us the tools we need to quietly spy on aliens from across the galaxy.

---

* I mean no disrespect to microbes when using this "dumb" versus "smart" distinction. Microbes do amazing things, like fermentation (thank you, microbes), and they can work collectively in remarkable ways. We should all be deeply in awe of microbes and fungi and forests. Still, for our discussion, it's useful to have a clean split between searching for the kinds of life that do not build technological civilizations and the kinds that do.

BIOSIGNATURES
How to find life from a distance

The nearest planet to Earth is the one orbiting Proxima Centauri (part of the triple-star system Alpha Centauri). It's more than four light-years away, and it would take more than eighty thousand years to travel there using the best spaceship tech we have now. That means we won't be landing on any exoplanets and searching them for life anytime soon. Thankfully, scientists are ridiculously stubborn and clever, which is the whole point of this book. With the new generation of telescopes they're building, a path has opened to search for life across all those trillions of miles. That is the true fruit, the true "Oh my god!" moment, emerging from the exoplanet revolution.

After thousands of years of asking, we'll be able to finally answer our alien-life question, and we're going to do it using *biosignatures*. We've been studying Earth's history for a long time now, and we've learned a lot about how life has shaped our planet's history. One of the most important things we've come to understand is that the sum total of Earth's life—its *biosphere*—has been a major player in our world's evolution for billions of years. This idea of a biosphere is going to be really important, so we're going unpack much more about it in a moment, but even eyeballing it quickly gives us some key questions. Do other planets have biospheres? Have they changed their planets as much as ours changed Earth? Most important, can we see that change from a distance? Can we find signatures of those biospheres in the light of alien worlds across interstellar distances?

To answer these questions, we're going to take a ride to the cutting edge of astrophysics. But to get there, we'll have to make a couple of stops along the way and pick up supplies (i.e., key ideas).

One of the best ways to detect an exoplanet is the "transit" method. We introduced this in the last chapter, but a wee reminder here won't hurt us. When a planet's orbit around its host star takes it between us and the star, the planet will block a tiny bit of the star's light. Here on Earth, we can use sensitive telescopes to detect that slight fading of the star's brightness. Detecting such a diminution of starlight is what lets us throw up our hands and yell, "Hooray, an exoplanet!" However, we can do something even more remarkable with a transit.

When a transit occurs, starlight will also pass through the planet's atmosphere (if it has one). As the starlight traverses the thin veil of planetary atmospheric gases, some of it will get captured by the alien air's atoms and molecules. This "absorption" of light is the key to detecting life across trillions of miles of deep space.

More than a century ago, physicists discovered an amazing link between light and matter. To understand what they found, we'll start by remembering that light is nothing more than waves of electromagnetic energy. The different colors of light correspond to different wavelengths of this radiation. As you'll remember, red colors are longer wavelengths, and blue colors are shorter ones.

Imagine that you pass a rainbow of light with all its colors through a box of pure hydrogen gas. Physicists started doing this kind of experiment in the late 1800s. When you look at the light coming out of the box, you see the same rainbow you'd put in, but there are some dark bands in the rainbow too. It looks as if something in the box of gas took a bite out of the rainbow at just a few specific colors. Some of the blue wavelengths are gone, and maybe a band in the red will be dark. The same thing happens if you use a box of sodium atoms, only now the dark bands are at different colors. In both cases, what happens is that atoms of hydrogen or sodium have *absorbed* some of the incoming light. Each element absorbs only specific wavelengths. The other wavelengths passed unmolested.

Hydrogen and sodium each have a kind of "spectral" fingerprint for absorbing light. In fact, every element has a different fingerprint of light absorption. Molecules do too. For astronomers these spectral fingerprints have been a godsend and a gold mine. It lets them gather light from distant objects, spread it apart into the component wavelengths, and then identify all the dark absorption lines from different elements and molecules. In this way, they can figure out the object's actual composition. To me that's a frackin' miracle.

OK, now on to planetary atmospheres.

By analyzing starlight that has traversed an exoplanet's atmosphere, astronomers can nail down exactly what the atmosphere is made of. If there are absorption lines of water in the starlight's spectrum, then there's water vapor in the planet's atmosphere. If there are absorption lines of carbon dioxide, then there's carbon dioxide in the planet's atmosphere. This process is called *atmospheric characterization*. It is even more of a scientific miracle, and I'm not letting us move on until we properly genuflect at its awesomeness.

Sitting here on Earth, an astronomer can figure out exactly what's in a planet's atmosphere even if that planet is tens, hundreds, or thousands of light-years away. Thanks to this radical new method, we can see exactly what's floating around in the atmospheres of planets that may never, ever be visited by a human. It is totally insane and amazing that we *Homo sapiens*, basically just a bunch of hairless monkeys, have figured out how to probe distant alien atmospheres. If nothing else, this achievement should make you a little bit proud of us as a species in spite of all the other horrible stuff we do.

Now let's get back to the story.

Atmospheric characterization is a big deal by itself because it will let us understand all kinds of planets in all kinds of new and important ways. We will, for example, be able to see what's happening in the outer layers of the hot Jupiter gas giants that are so hellishly close to their parent stars. Using atmospheric characterization, we can ask what kinds of chemicals form in an atmosphere that's hotter than a pizza oven. That will be great for understanding basic stuff about hot Jupiters, but what does atmospheric characterization have to do with life? Why will knowing what's in an exoplanet's atmosphere tell us what's living on its surface. To answer that question, let's go meet James Lovelock.

James Lovelock was a polymath. He was a genius at lots of stuff.

Born in 1919, he started as a chemist doing medical research in World War II. In the 1950s, he invented a cheap, portable device for detecting minute amounts of chemical contaminants in the air. The patent was so valuable, it made him rich and let him do whatever he wanted, including help governments that were just ramping up their space pursuits.

One of the government projects Lovelock signed on to was NASA's plan in the 1960s to send probes to Mars to look for microbial life. Lovelock, however, saw a huge flaw in this approach. All the mission plans assumed that life on Mars would be like life on Earth. If that assumption was wrong, then the probes would miss life even if it was right under their technological noses. Lovelock wanted a way to find life no matter what form it took. After a few years of work, he found that the answer was literally under his nose too. The solution was the very air he was breathing.

Lovelock realized that a planet's atmosphere was itself a kind of life detector. Earth's atmosphere has chemicals in it that would not exist on a lifeless world. Oxygen is a great example. Earth's air did not start off with oxygen in it. For the first two billion years, that element was basically absent.

How did our atmosphere suddenly end up being oxygen-rich? Life did it.

It was life that pumped the oxygen into the air after evolution invented a new form of photosynthesis. Sometime around two billion years ago, a kind of photosynthesis appeared that split water (H_2O) apart and spit the oxygen (the O) back into the air. Now the atmosphere is 21 percent oxygen. How this "great oxygenation event" occurred is one of the coolest stories in Earth science, but what matters is this: if all life disappeared tomorrow, the oxygen would also quickly disappear. The key idea here is that Earth's *biosphere*—the sum total of all life on the planet—keeps the atmosphere oxygen-rich. What's really happening is that life is keeping the atmosphere out of *chemical equilibrium*. That's a fancy way of saying that the biosphere continually pumps chemicals into the air that would not stay there for long on their own. That's how the chemical composition of the Earth's atmosphere tells us that Earth has a vibrant biosphere. Atmospheric oxygen, all by itself, is evidence that Earth is a living planet.

Now, finally, we can put all the pieces of this story together. From their new exoplanet atmospheric characterization methods, astronomers can get the chemical composition of an exoplanet's atmosphere. And what did Lovelock say we needed to see if a planet had life? The chemical composition of an exoplanet's atmosphere. That's it. That's the answer. If we can find the signatures of chemical disequilibrium in a planet's atmosphere, then we will have found evidence for life. We will have found a biosignature.

No matter how far that planet is away from us, biosignatures will be proof that the planet has a biosphere, proof that it is a living alien world. That, dear reader, is why we're standing at the edge of a revolutionary new era in the search for aliens. The discovery of exoplanets showed us where to look. The development of atmospheric characterization and biosignature science tells us what to look for. As the great Sherlock Holmes said, "The game is afoot!"

I could leave it there because . . . You know . . . Holy crap! We have everything we need to find alien life. But since a subtheme of this book has been those pesky standards of evidence, I want you to know that this adventure will have its share of danger. We will have to be super-careful in developing our target biosignatures. Maybe there are false positives out there. These would be all the ways a planet could fool us, producing what we think of as a biosignature without actually having a biosphere. For example, we already know at least one exotic way to produce oxygen in an atmosphere that has nothing to do with life. This can occur on those red-dwarf-star planets we met in the last chapter. Because these planets are so close to their star, water vapor atoms (H_2O) high in the planet's atmosphere can get punched apart by incoming starlight. The hydrogen atoms float off into space, but the oxygen atoms sink down and fill the atmosphere. We think we know how to account for these kinds of false positives, but there may be others out there that we haven't thought of yet. Care will have to be taken.

There will be other pitfalls and challenges that we will have to be ready for. But overcoming pitfalls and challenges is what gets scientists up in the morning. We will work them out. However, the challenges do mean we're playing the long game. This is a project whose progress is probably going to be measured in decades. That's OK. A few decades of being careful and clever is nothing compared to the two thousand five hundred years it took to get us where we are today, standing on the threshold of discovery, about to step across.

TECHNOSPHERES AND NOOSPHERES
When smart life goes boss

Life hijacks planets. Once it appears, life can take over, as it did on Earth, altering the fate of oceans, ice, air, and land. As we know, all that oxygen your lungs are enjoying so much is in the atmosphere only because life put it there. But it's not just oxygen. The root systems of plants can alter how much soil gets washed into the oceans. Ocean algae create tiny particles that climb high to the atmosphere, shaping rainfall patterns. The list of examples is long, but the thing you really need to know is that a sterile, lifeless Earth would look nothing like the one we live on.

The changes made by "dumb" life, like microbes, are ramped up on steroids for a technological civilization. So, if biospheres are important in shaping a planet's evolution, *technospheres* will be even more potent. They are the key to finding alien civilizations out there in the galaxy. They are the target of our NASA technosignatures research program.

In the previous chapter, we introduced the idea of a biosphere, the collective activity of all the life on a planet. Biospheres are a relatively new concept in science. It was first developed by Vladimir Vernadsky, the greatest scientist you've never heard of. Vernadsky was a Russian researcher working before and during the Communist era, which is why you've never heard of him. He was a genius at thinking about life and planets. He invented the sciences of geochemistry *and* biogeochemistry. For Vernadsky, life wasn't some green scruff that had no effect on its world. Instead, it was a dynamic force that shaped a planet's history.

In a series of lectures given in Paris in 1926, Vernadsky introduced the world to the modern concept of the biosphere. He detailed how life had to be a major player in planetary evolution, just as significant as the other geosystems—atmosphere, hydrosphere (water), cryosphere (ice), and lithosphere (land). Today, all our understanding of the Earth, including climate change, rests on the foundation Vernadsky laid down.

What does Vernadsky have to do with the search for life? We've already seen how the atmospheric characterization techniques astronomers recently invented allow them to search for biospheres by finding biosignatures. Looking for alien life really means looking for alien biospheres. Score one for Vernadsky. But he didn't stop there. In those same Paris lectures, Vernadsky pressed onward beyond the action of microbes, plants, and animals. The next step he laid out was thinking about how technology could be an even more potent force in hijacking a planet. This is how Vernadsky imagined the coming of *the noosphere*.

Noos (νοος, pronounced nó-os) is the ancient Greek word for "thought" or "mind." Even in 1926, long before climate change reared its head, Vernadsky could see the impact human civilization was having on the planet. He could also see how that impact was increasing. Eventually, he said, the biosphere would be overlaid with a "sphere of thought." This would be the collective impact of all human rational planning and technological development—the noosphere. These days, it's called the technosphere.

You don't have to look very far to find the technosphere we've built. Right now, as you read these words, there are invisible electromagnetic waves of thought passing through your body. I'm not talking about some kind of New Age telepathy. Instead, I'm talking about the wireless Internet. The Earth is now entirely surrounded by a shimmering sphere of thought, knowledge, and action embodied in our wireless technologies. And all this thinking, knowing, and acting has real-world consequences too. Human beings have reshaped the planet so completely that scientists now talk about the Earth entering a new geological era called the Anthropocene. We have, for example, colonized more than half the planet's surface for our uses. We move more nitrogen and phosphorus around the planet than natural forces. Our domesticated food animals weigh more than all the other mammals on Earth. And, of course, we've changed the planet's atmospheric chemistry and its climate. For better or for worse, the technosphere is here. We built it, and now we live in it. That is exactly the point when it comes to searching for alien civilizations.

If biospheres are the key to finding "dumb" alien life, then technospheres will be the path to finding "smart" aliens. Once a civilization begins harvesting enough energy and putting it to work to build a vibrant technosphere, it will have changed its planet enough to leave an indelible imprint. The surface will look different. The atmosphere will have different compounds in it. The space around it will be populated with machines (i.e., satellites) that never would be there without the technosphere. In other words, once a technosphere appears, the planet will begin to show *technosignatures* that can be seen from across interstellar and perhaps even intergalactic distances. Those technosignatures are things we humans will be able to see directly, but the possibility of their existence is something scientists have recognized only in the last decade or so. Ironically, we've become able to recognize them only because we've just now come to see our own technosphere as the planet-shifting force it is. That's how we've begun imagining what alien technospheres with a thousand- or million-year head start might look like.

TECHNOSIGNATURES
The day the Earth stood still-ish

Kids in a candy store. That's what we were thinking. It was Wednesday morning, September 26, 2018, and along with about fifty other astronomers, I found myself in a lecture hall at the famous Lunar and Planetary Institute in Houston, Texas. We'd all been called there by NASA to participate in a meeting titled "A Technosignatures Workshop." Apparently, Congress had just instructed the space agency to allocate $10 million to the search for alien civilizations. They wanted us to tell them how to spend it.

Federal funding for SETI had been hard to come by for quite a while, owing to how UFOs and images of little green men had been used to turn SETI into a political football back in the early 1990s. But the recent discovery of exoplanets had made astrobiology a major piece of NASA's efforts. While SETI was out in the cold, there was significant funding going toward the search for what I've been calling "dumb" microbial life on other planets. NASA had, for example, a number of projects aimed at detecting alien biospheres via their biosignatures. But for reasons that were out of NASA's control, the search for intelligent life was still getting no love. What made this NASA-sponsored meeting so exciting was the thought that maybe, just maybe, the Sun was about to rise on the search for "smart" life. NASA seemed ready to like the idea of technosignatures.

The term *technosignature* was coined by Jill Tarter, one of the great heroes of SETI. She began her career in the 1970s, and over the subsequent decades, she championed the search for cosmic life with courage, tenacity, and creativity. When the science of biosignatures was rising to become a serious research concern, Tarter saw that SETI in all its forms must be seen as part of that effort. Intelligent life is still life, after all. Tarter recognized that SETI had always been about finding technosignatures. Once she put that word in play, it became possible to see that a technosignature is just one very special kind of biosignature. That recognition is what helped get this meeting going.

When SETI began with Frank Drake's Project Ozma back in 1960, the emphasis had been on beacons. To detect a signal from a distant star, Drake and most subsequent SETI researchers had to assume that the aliens were concentrating their radio energy in a single direction. If the alien transmitter wasn't beamed but was sending energy in all directions, its power would have to be so high that it would basically almost be a star. That was asking a lot of our hypothetical aliens. Therefore, everyone looked for beacons, but that decision implied that the only signals we'd detect would be *purposeful* ones. The aliens would have to intentionally set up a beacon to beam radio waves into space and purposely announce their presence. This kind of search strategy had guided much of SETI for sixty years. Then the discovery of exoplanets opened up the search for "smart" life in new directions, including by using the radio telescopes that classic SETI had used.

Biosignatures are the *unintentional* imprint of a biosphere's presence on the light coming from a distant planet. It's not as though the microbes, forests, and animals making an alien biosphere are trying to announce their presence to the universe. The biosignatures just happen. The same is true of a technosphere. In other words, civilizations will create technosignatures regardless of whether or not they want to send out calling cards, simply by going about their civilization-building business. Moreover, as we joyfully discussed at the meeting, there are many ways for this kind of unintentional imprint to be produced. Pollution, city lights, reflections from solar panels, orbiting megastructures—the list was long and head-spinning. We were nerding out, big time.

The next three days were a wild ride. The first order of business was to survey the field. What kinds of ideas about what kinds of technosignatures had scientists already proposed? What kinds of telescope technologies existed now or were on the drawing board that could find those technosignatures? It soon became clear we had a lot to wade through. The most pressing question became: Given a basket of proposed technosignatures, which ones should we focus on?

A bright young scientist named Sofia Sheikh took this problem on. She digested the discussions and quickly gave us a proposal for how to sort technosignatures based on how likely they were to occur on an alien world, how long they'd last, and how easy they'd be to observe compared to other categories. Sheikh also suggested that we balance things like the ambiguity of a specific technosignature against the cost of mounting a search for it. Alien megastructures, like the Dyson spheres of chapter 2, might be cheap in terms of how much infrared telescope time was required, but it might also be true that any detections would be difficult to distinguish from natural astronomical sources, like big dusty stars. If so, then megastructures would score low on the balance of cost versus ambiguity. If we could find a different technosignature that scored higher, we should put our efforts into that one rather than into megastructures.

There were also fresh exciting possibilities in "classic" SETI's radio telescope strategies.. The use of artificial intelligence to automatically comb through zillions of hours of observations meant new ways to carry out "anomaly" searches. An anomaly is a signal that doesn't look like anything nature could produce on its own. This was a big focus of the Breakthrough Listen initiative that billionaire Yuri Milner had contributed $100 million to. Breakthrough Listen meant there was finally some real funding available to buy radio telescope time and get a comprehensive SETI effort started. Breakthrough Listen alone was reason to be excited for the future.

We discussed questions about technosignatures for hours and hours in the meeting. Then we discussed them for more hours over dinner. Afterward, we discussed them for a few hours over drinks and then in the hallways of the hotels we were staying at. By the time the meeting ended, we were all completely exhausted, but we were also completely exhilarated, as if we'd just spent a long weekend shredding impossibly big waves or skiing steep terrain on a fresh powder day. Most of all, we had the sense that a door was opening. Along with the already incredibly exciting possibility of biosignatures, the search for alien civilizations was also rising.

By the way, the $10 million never made it into the final congressional budget. Womp, womp, womp. It didn't matter. The snowball was rolling down the hill. Within a year, my colleagues and I would submit a proposal and be awarded NASA's first-ever research grant for atmospheric technosignatures. Soon a few other technosignature grants found success too. It was the dawn of a new era. That long-sought target—a well-supported scientific search for alien civilizations—was, hopefully, coming into sight.

ATTACK OF THE ALIEN MEGASTRUCTURES
Boyajian's star

In building the new science of technosignatures, the guiding question has become: What, exactly, are we going to look for? What kinds of technosignatures are folks in my science community focusing on? What do we think aliens will be doing, and how will we find them doing it?

Alien megastructures are the place to start this story. In the fall of 2011, a group of amateur astronomers was poring over data from the Kepler Space Telescope, designed to be a wholesale exoplanet-finding mission. Kepler could stare at thousands of stars at once, looking for the moment when a planet passed between us and its host sun. When this happened, it would register a small dip in the star's brightness, and voilà, another strange new world would get added to its long list of discoveries.

Most of the time, those dips in star brightness (transits) fit exactly what was expected from mathematical models of a planet passing in front of a star. This regularity allowed NASA to build AI that would look at the data before anyone else did and flag the stars that looked as though they had planets. But sometimes Kepler saw things so strange that the AI got an AI version of indigestion. In that case, the computer immediately puked the data back to its human creators in a "you figure it out" kind of way. This happened enough times that NASA created a citizen-science website called Planet Hunters[1] that let interested nonscientists look over the space oddities. That's how the star affectionally known as KIC 8462852 made global headlines.

Normal transits just look like a U-shaped dip in the brightness of the star that repeats once every orbit. The planet swings around the star. It passes between us and the star, blocking out some light in a way that looks like a U on a graph of brightness versus time.

However, instead of those nice, regularly timed U-shaped dips, the light from star KIC 8462852 looked like the grin of a dragon with dental problems. The light would be sharply blocked for a short time, creating a deep V shape in the brightness. And then nothing, until a little later, three short, sharp dips of different depth and then nothing again for a long time. This wasn't like anything a planet could create. Something very strange was happening around the star. Soon the crazy data was brought before Tabatha Boyajian, a young astronomer working with the Citizen Science project. It didn't make sense to her, either. So she did what any self-respecting scientist would do. She started writing a paper. As part of that effort, she spoke with Jason Wright, a professor at Penn State University (and a member of our NASA technosignature team). Intrigued, Wright went on to write a paper that tried to systematically identify all the possible kinds of stuff that might be passing in front of KIC 8462852 to make the transit signature so weird. How about a cloud of orbiting comets? Maybe. What about chunks of a broken-up moon? Could be. Might it be a series of dust clouds? Yeah, that's possible. Oh, and how about a swarm of orbiting alien megastructures?

Cue the record scratch.

Yeah, alien megastructures. Wright decided to include orbiting alien megastructures as one possible explanation in his paper. Why would he do this, especially knowing he might catch some serious grief for making such a suggestion? Was he just being cheeky? Was he looking for a fight? Was he looking to get on TV? The answer is much simpler. At the beginning of this search for an explanation of an anomalous astronomical signal, these megastructures were a very old idea with a very good pedigree.

The possibility that an advanced alien civilization would surround its star with giant solar collectors goes all the way back to the early years of SETI and the idea of Dyson spheres. Ever since those initial papers, Dyson spheres and Dyson swarms have become an essential idea for scientists thinking about the observable imprints— the technosignatures—of alien civilizations. In 2005 astronomer Luc Arnold carried out a detailed study of exactly how orbiting alien megastructures, like pieces of a Dyson swarm, would appear in telescopic observations. Arnold began his calculation by considering different shapes of alien megastructures—triangles, squares, rectangles, and so on. The "mega" part of his thinking was simply the fact that such machines would need to be as large as a planet. The "alien" part was the fact that nature, as far as we know, doesn't make planet-size triangles. By calculating how the structures would block light as they passed in front of the star, Arnold assembled a catalog of alien megastructure transits. Some of Arnold's theoretical light curves looked suggestive of what Boyajian saw in the KIC 8462852 data. Wright pointed this out in his paper.

Boyajian obviously wasn't claiming such a discovery, and Wright was merely saying it should be on the list of possibilities. When Wright and Boyajian put in a telescope proposal to look for signatures of a Dyson swarm around KIC 8462852 the press got hold of the story and soon the term alien megastructures was sitting there in big, bold headlines all over the planet. The reporting on KIC 8462852—or Boyajian's star, as it came to be known—was the first public inkling that something had changed in the search for alien civilizations.

Eventually, though, after much hard work and new observations, astronomers found that the best explanation was clouds of dust, not planet-size machines. Yeah, I know, what a bummer. Clouds of dust aren't as exciting as alien megastructures—unless, of course, you're an astronomer who studies clouds of dust, and then, let me tell you, they're pretty damn exciting. But the disappointment is not really the point of this story. What really came out of the Boyajian's star episode were two really important recognitions. The first was that the exoplanet revolution had totally rewritten the rules about looking for alien civilizations. The second was that the damaging giggle factor was finally on the wane. It was becoming OK to include technosignatures as a possibility in peer-reviewed scientific papers.

Wright's suggestion of a possible alien origin for the mysterious data was reasonable until proven otherwise. It was a milestone in the history of the modern search for life and intelligence. Just as important was the fact that astronomers had seen the exact kind of signature we'd expect for one of the most anticipated forms of alien super technology. Even if Boyajian's star didn't have them, the possibility that some other star someday might became that much more plausible.

POLLUTION, CITY LIGHTS, AND GLINT
What alien skies can tell us about aliens

We finally have telescope technologies powerful enough for the search (the James Webb Space Telescope and beyond), and we finally know exactly where to look in our search (exoplanets), but what exactly are we searching for? One strategy is to just look for weird stuff, or more formally, *anomalies*. This is the very exciting focus of a lot of radio SETI these days, as researchers use advances in artificial intelligence to comb through oceans of observations and search for signals that don't look like anything nature produces. Another route, the one my colleagues on the NASA grant are taking, is to unpack the ways any civilization anywhere might evolve. We need to systematically imagine how technology evolves and figure out its observable imprints, if any. In particular, we need to know what civilizations in general will do to their planets. Luckily, the history of humans and Earth gives us one likely answer— pollute them.

As strange as it might seem, looking for pollution in the atmospheres of distant worlds may be the fastest way to find a distant civilization. Right there in the definition of industrial civilizations is the important word *industry*. What's industry? It's the use of energy to make stuff. The more *advanced* (beyond your earliest tools) the stuff you make becomes, the less it looks like nature. If some of that advanced, unnatural stuff can be seen from a hundred light-years away, then *boom!*—you have a technosignature.

Maybe that sounds too abstract. Let's bring the discussion down to Earth, literally. Chlorofluorocarbons (CFCs) are a class of molecules invented by chemists back in the 1920s. *Invent* is the key word here, because CFCs don't occur naturally. As a mix of the elements chlorine, fluorine, and carbon, (I know, duh), CFCs have remarkable properties that make them perfect for industrial applications. Their response to heat makes them great for air conditioners. Their reaction to pressure makes them ideal in spray cans. Thus, billions of tons of CFCs found their way into factories, offices, and homes around the world. However, because of all that industrial production, millions of tons of CFCs also found their way into the atmosphere. That's when scientists discovered that CFCs were eating Earth's ozone layer.

Oops.

This was very bad news because the ozone layer is what protects us from cancer-causing solar radiation. Once people realized what was going on, they got together, held hands, sang "Kumbaya," and then severely curtailed the production of CFCs. It was a very big deal, and our response became the model for dealing with greenhouse gases and climate change. These days, there are still CFCs in our atmosphere but at much lower levels.

This tale of industry, air, and chemistry shows us two things that are really important when it comes to hunting aliens. First, chemicals like CFCs exist only because of technology. Second, an industrial civilization can pump huge quantities of those chemicals into their planet's atmosphere either on purpose or by mistake. For us and CFCs, it was by mistake, but a civilization might intentionally add chemicals to their atmosphere for all kinds of reasons (like terraforming, which we'll explore later).

Now, here's the kicker. Some of the atmospheric chemicals a civilization puts into the atmosphere might be detectable across space. I know this for a fact because it was one of the first results to come from our NASA technosignature research group. We did it using CFCs. Led by Jacob Haqq-Misra and Ravi Kopparapu, our team built a mathematical model of an Earth-like planet orbiting a star tens of light-years away. Then we put the same levels of CFCs in its atmosphere that we have in ours right now. Finally, we "observed" that exoplanet with a mathematical representation of the James Webb Space Telescope. Our results showed that, given some assumptions, using the JWST for just a few weeks was enough to detect CFCs in our simulated inhabited alien planet. If this had been a real planet, we would have found conclusive proof for the existence of an alien civilization. We were pretty proud of, and pretty excited by, these results.

Our results were a kind of milestone for technosignature science. For the first time, scientists (our team) had shown that industrial technology *on the level Earth has now* (CFC production) could be detected using *telescope technology the Earth has now* (the JWST). That translates into a simple conclusion: scientific alien hunting on alien planets is now a thing. This doesn't tell us there *are* industrial chemicals in any exoplanets' atmosphere, but it does tell us that if there were and they were at the right levels, then we could find them.

Chemicals in an atmosphere are not the only way distant planets might reveal that they're hosting a civilization. Check out a picture of Earth at night, and you'll see how all the cities and the roads connecting them are beautifully lit up like luminous spiderwebs. An alien visitor who reached Earth's orbit could look at the nightside of our world and instantly know that Earth had a high-tech species. Could we do the same by looking for city lights on distant planets? In other words, could artificial illumination be a technosignature? The answer is a definitive yes, thanks to all those advances in telescope technology I keep going on about.

Artificial lighting sources produce strong spectral imprints that could be seen from an exoplanet. Technologies like the sodium or halogen lamps we use today have a collective global effect. The signature of these artificial lights will be encoded in the light from any alien planet with a civilization that uses them. In 2021 Thomas Beatty showed how these technosignatures could be detected using telescopes on the drawing board right now.[2] There was a caveat to his result, though. Seeing alien artificial illumination with the telescopes we have now might require a planet with a lot more city lights than Earth. Our best shot would be seeing planets with huge world-spanning cities like Coruscant from *Star Wars* or Trantor from Isaac Asimov's Foundation series. Scientists call these kinds of planets an *ecumenopolis* (i.e., a city-world). (And I'll give you a buck if you can drop *ecumenopolis* smoothly into a conversation today.)

As our telescope technology gets better over the coming decades, we'll be able to do more than just look for imprints of technology hiding in planet light. The new generation of ground-based telescopes being constructed right now will have thirty-meter (hundred-foot) mirrors. That's almost three times bigger than today's telescopes. By using these beasts and advanced techniques in analyzing light reflected from an exoplanet, astronomers may be able to make low-resolution images of its surface.[3] This will be a truly revolutionary advance when it happens. The first thing we might do with monster telescopes and the new technique would be to look directly for bright regions on a planet's night side. We could also map local sources of heat to track sites of industry where energy is being harnessed to do the work of maintaining the civilization. If mapping an alien planet in this way seems radical to you, your instincts are right. If you're under the age of fifty, there's also a decent chance you'll be alive to see it happen.

The last planet-based technosignatures worth mentioning are solar collectors. In 2017, Manasvi Lingam and Avi Loeb revealed how civilizations using solar panels on planetary scales create clearly detectable technosignatures.[4] Lingam is a really gifted theorist with an encyclopedic knowledge of the scientific literature. When sunlight hits a solar collector, some light is absorbed, and some gets bounced back into space. Think of the glint off any reflective object. The light that's reflected carries an imprint of its encounter with the collector. Lingam and Loeb showed that this imprint would be there no matter what kind of material the aliens used to harvest solar radiation. Nature provides only certain kinds of elements, like silicon, that respond in the right way to turn sunlight into electricity. Lingam and Loeb called the signature that solar collectors leave on reflected light a UV edge. It's basically a huge jump in the reflectivity happening right around the ultraviolet region of the spectrum.

What's particularly cool about this result is that astrobiologists have known for a while that our biosphere imposes a strong "red edge" in Earth's spectrum. By observing Earth with our own spacecraft, scientists discovered that there is a big jump in how Earth reflects sunlight at the boundary between green and red wavelengths. This red edge is created by sunlight bouncing off all the leaves on all the trees in all the forests on our world. Something similar should happen on any planet where photosynthesis is happening in a big way. The red edge has been a favorite potential biosignature for a while. Now, thanks to Lingam and Loeb, the UV edge may be one of our best potential technosignatures.

• • •

SOLAR SYSTEM ARTIFACTS
Did you leave these?

On July 20, 1969, Neil Armstrong and Buzz Aldrin safely guided their lunar lander onto the surface of the Moon. They planted a flag, deployed some scientific instruments, and then blasted back into space, leaving the descent stage of the lander (its lower half) on the surface. They were on the Moon for just about twenty-two hours. Those hours were certainly momentous, but they pale in comparison to how long the stuff they left will remain sitting there on the Moon. Without the wind and rains to erode them, the artifacts Armstrong and Aldrin left will stay intact and discoverable for possibly millions of years or more.

This remarkable timescale raises an equally remarkable question. If our stuff is going to remain intact on the Moon for so long, what about stuff left by someone, or something, passing through on a visit? In simpler terms, if aliens passing through the solar system a long time ago left equipment on the Moon, would we still be able to find it?

This is the question that's raised by an entirely separate branch of technosignature studies called solar system SETI or artifact SETI. The basic idea is as simple as it is stunning. As solar system exploration reaches its next stage of maturation, we should spend some of that time looking for stuff left behind by aliens. To see exactly what solar system SETI researchers are talking about, let's start out with the kinds of searches that are just becoming possible. The Moon is a good place to start.

In the decades since the Apollo program, there have been many robot orbiters sent to the Moon to image its surface. The pictures have become so good that we now have an almost complete map of the lunar terrain, down to a resolution of about a meter. With that kind of clarity, we could, in principle, search the entire 14.6 million square miles of the Moon for evidence of technologies left behind by someone other than us. If you did this by hand, it would take a lifetime to finish. No one is that patient. Thanks to artificial intelligence, however, you can get a computer to do the work for you. Using machine learning, a computer can be trained to look for anomalies in those surface images. Here, anomalies are stuff that doesn't look like rocks. In 2020 a team of researchers from NASA's Jet Propulsion Lab did exactly this for a much smaller region around the Apollo 11 site. The AI had no problem picking out the lander in the relatively small target region the team directed the AI to look at. The next step, if the team can get funding, would be to do the same search using images for the entire Moon.

Beyond searching the surface of moons and planets for alien technological artifacts, we could also look for alien machines sailing around interplanetary space. The problem here is the constant gravitational tug of the planets as they swing around the Sun. Over time these little tugs and pulls will destabilize an artifact's orbital motion. After hundreds of thousands to millions of years, most orbiting artifacts left by aliens would get dragged into the Sun or have crashed onto a planet. To make this approach work, scientists need to identify orbits around the Sun or planets that are stable for billions of years. Those long timescales are needed because there is no way of knowing when our hypothetical alien visitors actually visited. If they passed through when the dinosaurs ruled, a hundred million years ago, they'd have had to place their probes in just the right kind of orbit for us to still find them.

Now let's deal with the real brain-scrambling question underlying all of this: Could aliens *really* have visited our solar system? I am inclined to think that this is a longer shot than looking for technosignatures from alien planets themselves. If you want to find aliens, go look where they live (exoplanets), not some backwater planet like Earth.

That said, I first seriously encountered the idea that the solar system might be home to alien artifacts during a presentation at a technosignatures meeting in 2020. The speaker was talking about *lurkers*, the term given to hypothetical probes sent by another civilization to observe Earth. At that moment, it hit me that if we were to find a planet that had biosignatures, we would, as soon as we were able, send something to lurk on them. We'd send a robot across interstellar space to observe the biosphere-hosting planet and send back data. If everything worked, we'd send another one, and then another and another. I ain't gonna lie; the thought gave me the chills. If there was ever a way the UFO-as-aliens thing could make sense to me, this was it. Not triangle-headed visitors looking to beam up some cattle, but machines sent across the impossible void to collect data. Before I signed on to this idea, though, I would have lots of questions about why these probes seemed to be advanced enough to hide from us while simultaneously being so bad at cloaking that we always almost see them (the high-beam argument from chapter 4). And that's just the *first* of my questions.

However, you don't have to believe in UFOs as aliens to still think solar system SETI is worthwhile. The Sun and its planets are more than four billion years old. Artifacts sent to the solar system or left by voyagers passing through may have arrived here at any time during those eons. They could be long dead and inoperable but still be detectable. Recent or ancient, no matter how you think about it, solar system SETI artifacts represent a viable stream of technosignature research. Thanks to rapid advances in our own technology (AI and interplanetary exploration), the time to begin these searches in earnest has finally arrived.

WAS 'OUMUAMUA AN ALIEN PROBE?
You have a visitor

The strange object had been in the darkness of deep space for eons, quietly crossing the long cold distance between stars. Now the light of Sol, home star of Earth, was growing brighter as it crossed the orbit of Neptune and into the domains of our solar system. It wouldn't be staying long, though. Once humanity recognized it had a galactic visitor, there would be only a few months to study the interloper before it whipped past the Sun and back out into the black forever.

This is how *Rendezvous with Rama*, the 1973 science-fiction novel by Arthur C. Clarke, begins. A miles-long, cylindrically shaped alien craft hurtles silently through the solar system on an unknown mission to an unknown destination. First misidentified as an interstellar comet, humans must scramble to send a ship to investigate. Thirty-five years later, life would imitate art when 'Oumuamua suddenly appeared at the edge of our space.

The Panoramic Survey Telescope and Rapid Response System, or PAN-STARRS telescope, sits high atop Haleakalā volcano on the island of Maui in Hawaii. Run by the University of Hawaii, it's designed to continually sweep the sky, looking for transients—celestial objects that change rapidly or appear and then rapidly disappear. On the night of October 19, 2017, the PAN-STARRS telescope discovered a transient like no other seen before. The object, designated 1I/2017 U1, wasn't a comet or an asteroid swinging around the Sun for yet another orbit. Based on its speed and motion, 1I/2017 U1 was clearly just passing through. It wasn't part of the solar system at all. It was a visitor.

The object discovered that day in October was the first true interstellar interloper ever seen in the solar system, and it was big news. At first, some astronomers thought to name the object Rama, but the consensus soon landed on 'Oumuamua, meaning "scout" in Hawaiian. Within a day of 'Oumuamua's appearance, the astronomical community went into overdrive. Astronomers raced to make as many observations as possible before it disappeared back into the void between the stars. The main goal of the researchers' efforts was to answer the most basic of all questions. What was it? Theoretical studies had long predicted that comets and asteroids could easily get flung out of a solar system and into interstellar space. The main culprit for the ejection would be a close encounter with a Jupiter-size planet. As the data came in, however, 'Oumuamua seemed to resist easy classification.

The first oddity was the shape. By carefully monitoring the reflected light from 'Oumuamua, astronomers deduced that it had an extreme aspect ratio. This is a fancy way of saying it was either long and thin like a cigar or flat and round like a pancake. It's not easy for nature to create such shapes, and no object in the solar system had ever been seen that looked like 'Oumuamua.

The second was its motion. When a "dead" object, like an asteroid, orbits the Sun, the only accelerations (changes in velocity) that can occur come from gravity. The Sun's gravity pulls it toward the Sun. The planets will also exert their own tugs. With a good computer, you can usually track all these gravitational contributions to an object's motion and predict all its accelerations. If you see more acceleration than can be accounted for by gravity, then it must mean the object isn't dead. Something must be happening on its surface. This often occurs on comets, which are basically mountain-size dirty ice balls. When comets swing close to the Sun, they warm up and some of the ice gets vaporized. Jets of steam begin busting through the comet's surface. Each jet acts like a little rocket motor pushing the comet this way or that. Such nongravitational accelerations are always accompanied by detectable clouds of water vapor and other compounds ejected by the jets.

Astronomers tracking 'Oumuamua saw a steady nongravitational acceleration. This should have led to the easy conclusion that 'Oumuamua was, indeed, the core of a comet ejected from some distant star system. However, when astronomers looked for gas or vapor jets, they couldn't find any. If there were jets, they were invisible.

So 'Oumuamua presented astronomers with a dilemma. They had seen something new in the sky, but they could not completely explain its properties. This is not that radical a situation for them. Astronomers often see stuff they don't understand. Only after exhausting all of the simple, easy, understandable explanations will they move on to the category best labeled "Holy shit!" That's where the alien spacecraft live.

Alien spacecraft were exactly the possibility Avi Loeb was willing to consider in the work he did with Shmuel Bialy. Loeb was the chair of the Harvard Astronomy Department when he and Bialy wrote papers that merely suggested that light sails might be a possible explanation for 'Oumuamua. As we discussed in chapter 4, light sails are an idea that's been bouncing around for interplanetary propulsion for a long time, and we have even tested a few in Earth orbit. For Loeb and Bialy, it was worth considering the possibility that 'Oumuamua was exactly this kind of machine launched by another species and passing through our solar system. It was the kind of suggestion that wasn't so different from what happened with Boyajian's star, its crazy transit signature, and the suggestions about alien megastructures. The difference in this case, and the cause of much consternation, was that once 'Oumuamua was gone, it really was gone. Within a year it was so far away that even our best telescopes couldn't see it. We would never be hearing from 'Oumuamua ever again.

In the absence of new data, all that was left was the old data. Based on the information on hand, most of the scientific community thinks 'Oumuamua was the burnt-out core of an extrasolar comet. There have been a variety of explanations offered for both the object's shape and its nongravitational accelerations. Some researchers, for example, have argued that 'Oumuamua was made from nitrogen ice rather than the more familiar water or carbon dioxide (dry ice). Unlike water or carbon dioxide, nitrogen gas is really hard to detect, so maybe that's the solution. Loeb and Bialy, however, argue that comets can't form from nitrogen ice. In general, Loeb continues to posit that all the "natural" explanations fail to account for 'Oumuamua's properties. For him, the best and most reasonable explanation is that it was indeed an alien artifact wandering through our backyard.

Given that 'Oumuamua is beyond our reach (though there have been proposals to send a probe to catch it), there is no way to definitely answer the question of its nature. For me, given the lack of new information, those standards of evidence I keep talking about make the alien hypothesis a less plausible one. We simply don't have enough data to conclusively conclude anything about 'Oumuamua. That means, while the alien light-sail possibility can't be ruled out, it shouldn't be at the top of the list, either. It's a question mark that will remain a question mark.

However, at this point, 'Oumuamua is no longer the point. We have already discovered another interstellar visitor passing through the solar system. Called 2I/Borisov, it was confirmed to be an extrasolar comet. If there have been two, there will doubtless be others. Now that we know, we will be watching. Perhaps the next visitor, or the one after that, will be an artifact of intelligence, one that might truly earn the name Rama.

TERRAFORMING

How to engineer a habitable planet

Good planets are hard to find. Whether it's human beings trying to find new worlds to settle or aliens looking for a replacement home, most planets just don't fit the bill. Many, like Mercury in our solar system, won't have atmospheres at all. And the planets with atmospheres probably won't have the gases your species needs to breathe. What's an advanced spacefaring civilization to do? Engineer the new world to make it just like home.

This process, called terraforming, is a staple of science fiction, appearing in movies and TV shows like *Aliens*, *Star Trek*, and *Firefly* and video games like *Horizon Zero Dawn* and *The Outer Worlds*. Elon Musk has been pretty vocal about terraforming Mars, and it does seem like something a sufficiently advanced civilization could pull off. If that's the case, the effects of terraforming might be obvious enough to be a big bright blinking technosignature.

THE COSMIC STAKEOUT

But to understand how we might find intelligent life by looking for its planet-scale engineering, we first have to understand how terraforming might work.

Let's say you have a crappy planet, and you want to turn it into a paradise. What do you need to do? What steps and what technologies do you need to effect the transformation?

Does your planet have no atmosphere?

Sounds like you have a real "starter" planet on your hands. First thing you'll want to do is find some comets and redirect their orbits so they crash onto the planet. Comets are full of things like carbon dioxide (CO_2) and water (H_2O), both of which you'll need to build up a nice blanket of gases around your world. CO_2 is a great greenhouse gas too, so you'll need it for getting to comfortable daytime temperatures. Water can be easily broken up to liberate oxygen, which can be used for breathing if you're into that sort of thing. Obviously, you don't want to hang around on the surface during this "drop comets on the planet" apocalypse phase of terraforming, but once the smoke clears (after a few centuries or so), you should be good to go.

You might also be able to get some of what you need from frozen reserves on the planet itself. There can be a lot of what planetary physicists call *volatiles*, which are good for atmosphere building, buried deep within a world. Volatiles are things like nitrogen, carbon dioxide, ammonia, and methane, which become gases at relatively low temperature (i.e., room temperature). Methane and ammonia would be terrible for humans (the former smells like fart gas and the latter eats your lungs), but they might be the scent of home cookin' for aliens. It is noteworthy that plans to terraform Mars often depend on melting CO_2 found on its extensive ice caps. Unfortunately, new research indicates that there may not be enough CO_2 locked up on Mars to get the terraforming job done.

Does your planet have an atmosphere, but not one that is useful to you (it's a lemon)?

If your planet already has atmosphere, but it's the wrong kind, you may still have to follow the steps above. Terraforming is about getting the atmosphere and the climate you want. If the planet is like Mars, then there will be "air" but too little of it to keep the surface warm. There also won't be enough oxygen to make our kind of life possible. So you'll probably have to look for stuff to melt and then vaporize to get it into the atmosphere. If that's not there, you'll need to do the comet thing.

For a planet like Venus, however, the story is reversed. Venus has *too much* atmosphere, and it's all CO_2. Terraforming a planet like Venus would probably require putting a giant sunshade in orbit to block sunlight. That would cool the CO_2 and condense it out of the atmosphere. What would be left would be a kind of foamy dry ice covering the surface. All that CO_2 would then need to be scooped up and buried. Snowplowing an entire planet is, obviously, not a task for the technologically fainthearted. Only a pretty advanced species could pull it off.

If you are being forced to engineer an atmosphere then you're going to need to genetically engineer your way to a biosphere too. If you want a garden world, you need to plant a garden. An inhabited planet means a rich biosphere with a zillion different kinds of species living together (and eating one another). You probably won't be able to just import the flora and fauna from your home planet, so you'll need to genetically modify some microbes to help kick-start the biosphere on your new world. Once these get going, they can also help cycle the right kind of elements into and out of the atmosphere. This is what happens on Earth, where oxygen in the atmosphere comes from life's activity. That's the model you want to follow.

That's it. Pretty easy, huh? Just get the process going, wait a few centuries (or millennia) while making necessary tweaks, and voilà, you'll be ready to inhabit your new world.

THE COSMIC STAKEOUT

So how can terraforming become a technosignature? One possibility that Jill Tarter suggested is finding a string of identical planets orbiting a distant star. If you look at the rocky planets in our own solar system, they are very different from one another. Mercury has no atmosphere. Venus has too much atmosphere and a runaway greenhouse effect. Earth has an atmosphere and oceans. Mars is a dry frozen desert with a thin atmosphere. Given that disparity, imagine that we found another solar system with four rocky planets and all of them had the exact same atmospheres and climates. Imagine that they all had 21 percent oxygen and average surface temperatures of 60° Fahrenheit. Such identical conditions would be unlikely to occur on their own and would be an indication that the planets had been purposely engineered to create a specific set of conditions.

Another way terraforming might lead to technosignatures would be to catch aliens in the act. If we wanted to terraform Mars, for example, we could pump artificial greenhouse chemicals into the atmosphere to warm the planet up. Remember, we already have a bunch of these in the form of CFCs. CFCs are much better at making the greenhouse effect than CO_2, and so they might be the kind of thing we might see in the atmosphere of a planet being terraformed. Unfortunately, no one really knows if terraforming is possible.

Maybe planetary climates just can't be engineered this way. Maybe, even if you could build the atmosphere you want, it would go unstable and collapse after a century or so. But if we survive our own little experiment in planetary engineering—i.e., climate change—then we might get a chance to answer some of these questions on Mars. Even better, terraforming seems like such a basic idea that we expect it's the kind of thing that any spacefaring species might want to try.

There you go. Megastructures, atmospheric pollution, city lights, worlds covered in solar panels, and terraforming. In this chapter, we've traversed a broad landscape of possible technosignatures emanating from imagined other worlds. In the process, we've gotten a handle on how technosignatures and biosignatures get tied back to life as it transforms the planets it inhabits.

The key point here is that life's not some frail little bunny hiding in the shadows of its big bad planet. Instead, it's got muscle. It's got power, and that power can reshape a world so completely that we can see the consequences even on our humble abode, Earth.

These are not just ideas. Searching for these signatures of extraterrestrial life is something we can do and are doing right now with the JWST. The Webb telescope, however, lives at the inner edge of the capacities we need to really get going. It's going to be the machines that are in construction right now, like a project named the Extremely Large Telescope, that are going to take the fight to its next level. Even better are the ones that are the drawing board now, like NASA's Habitable Worlds Observatory, which will be the successor to the JWST. Scheduled to be launched in 2040 or so, it's being designed with life detection as its highest priority. I may be old and a bit loopy by that time, but I'm totally sure many of you won't be, which means you will have a ringside seat as the data comes in.

It's gonna be a hell of a show.

CHAPTER 7

Do Aliens Do It Too?

WHAT WILL WE FIND WHEN WE FIND ALIENS?

Was this the first chapter you opened when you picked up the book?

I'm not judging, I'm just saying . . .

And what I'm saying is this. Of course, you want to know how aliens have sex. We all do. Really what we want is to know how aliens do *anything*. That's because anything aliens do must be so, well, alien. Given how vastly different their planets and histories will be, is there anything we can say about the aliens themselves that won't just be made up? Can we use science to guide our thinking about what aliens look like, what they do, and how they do it?

If you look at Earth, the variety of biological forms and functions are stunning, beautiful, and insanely varied. Life on this planet runs the gamut from amoebas to octopuses to bald eagles to humans with computers in their pockets. Shouldn't we expect alien worlds to be just as, or even more, varied? Looking at Hollywood aliens, we can also see how limited our imaginations (or their budgets) have been in picturing such variation. Most movie and TV aliens are just humans with prosthetic attachments glued to their foreheads or ears (I'm looking at you, Mr. Spock). Even many of the most compelling Hollywood aliens, like those in the movie *Alien*, still have our basic body plan. So, beyond just making stuff up and not doing a very good job of it, let's see if science can do a better job at guiding our expectations for alien physiology and behavior. The key to success will depend on navigating the intersection of physics, chemistry, and biological evolution.

So let's start with the chemistry, and ask ourselves what makes carbon so dang special.

BEYOND CARBON-BASED LIFE?
The molecule of love

"What about non-carbon-based life?"

That's often one of the first things people ask me when I give a talk on astrobiology. It's an excellent question, and I am always excited to talk about it. The thing is, astronomers and biologists have thought for lifetimes about which atoms life might choose for its building blocks. While there are some alternatives that are cool to consider, there's also a really good reason why you and I and every other living thing on Earth are made with carbon. Those reasons also favor a universe full of carbon-based aliens.

It all comes down to a simple fact. Carbon really likes to bond. Now, I'm not saying it's got no judgment. Carbon won't hook up with just anybody like, say, fluorine does, for god's sake (which is why fluorine is so dangerously corrosive). What makes carbon unique is its ability to form stable, flexible bonds with key elements like oxygen, nitrogen, phosphorus, sulfur, and even itself. Strong, flexible bonds make carbon the perfect candidate for building a biochemistry.

For an element to be a good basis for life, it must combine (bond) easily and in the right way with other atoms. Any form of biochemistry needs to be based on making lots of fast reactions that can build huge varieties of complex molecules. These molecules must also be modular, swapping pieces in and out of bigger molecular machines that do everything from making muscle tissue to being useful in information storage.

Carbon has all these traits in spades. Even so, to really understand what makes it so potent as a basis for life, and to see what alternatives do exist out there, you have to look at the periodic table of elements. Sorry.

Like many people, I grew up hating chemistry. All that memorization and the endless balancing of electrons in chemical formulas. I failed my General Chemistry class back in college and had to take it again. Over the years, however, as my research in physics turned into research in exoplanets and astrobiology, I've come to see chemistry for the beautiful and ornate miracle of nature that it is. Nothing exemplifies that beauty and power more than the periodic table. All those neatly ordered rows and columns are trying to tell us about a wonderful secret, a hidden order in nature. The columns with their different colors and the rows with their big gaps like a smile with a bunch of teeth missing... What's the message?

The structure in the periodic table reflects a profound architecture in the elements that make up the material world. More important, it's an order that holds across the *entire universe*. No matter where we look, no matter how far out we look, we always find the same elements with the same properties. We can tell this because we find the same elemental light signatures in distant galaxies as we find in Earth-bound labs. The periodic table of elements is truly universal. Those universal properties of the universal elements tell us a lot about the kind of biochemistries that alien life can build. For example, on the rightmost column of the periodic table, you'll find the *noble gases*, like helium, neon, and argon. These guys rarely form bonds with other elements. They don't like to mingle with the riffraff. Why won't the noble gases bind easily with other elements? It's the arrangement of their outer electrons orbiting their nuclei. Those electrons keep the noble gases noble. Without going into details, the location of an atom's electrons relative to its nucleus tells you a lot about its chemical proclivities. It's why you can forget trying to find krypton-based life.

Now let's jump left to the column topped by carbon. The promiscuity, flexibility, and resilience that make carbon so useful for life come from chemical pairings made possible by four outer electrons, all in the right place to make for easy chemical bonds. Every other element in that column of the periodic table also has four electrons in the right place (in terms of atomic structure), just like carbon. The next element below carbon is silicon. That placement is why it's always silicon that people talk about for non-carbon-based life. Because silicon has an electron arrangement like carbon's, it also will be able to form similar kinds of strong, flexible chemical bonds with other elements (including with itself). However, the periodic table also shows us why silicon will present some real challenges for life.

One problem for silicon is that it just isn't as friendly as carbon. Even though it has the four bond-available electrons, it's a bigger, heavier atom. That means there are fewer other elements it can make stable bonds with. This becomes truer the farther down the column you get. The relatively limited kinds of chemical pairings available to silicon translates into limitations in the kinds of molecular machines it can build. This matters when it comes to making *polymers*, which are the long, stringy chains of atoms from chapter 5. Carbon polymers do all kinds of crazy things, like fold themselves into big complicated balls of string. The specific shapes these polymer balls take give specific proteins their specific functions. For example, the protein hemoglobin carries oxygen around in your blood with an efficiency that's only possible because of the crazy but very exact way it folds on itself. That folding is possible only because of the carbon bonds in hemoglobin's chainlike polymer molecule. Silicon simply can't be used to make these kinds of long, flexible chains, because its bonds are more brittle.

DO ALIENS DO IT TOO?

Also, silicon can't swim very well. Carbon compounds work a lot of their biochemical gymnastics by floating around in water. But silicon chemistry is unstable in aqueous environments. The compounds made from silicon tend to just fall apart in water. Instead, silicon tends to work its pairings in solid form (i.e., crystals). If all silicon biochemistry had to be carried out in solids rather than liquids, that could severely limit its reach. Consider, for example, what that means for getting rid of waste, for breathing, peeing, and pooping. In a silicon-based creature, all these activities would be rigid and take a lot more energy. Silicon animals would literally have to breathe sand and shit bricks.

This doesn't mean life couldn't build itself with silicon. It may be that silicon life can only form in cold environments, where the pace of biochemistry would be a lot slower. Maybe it exists in places like Titan, the biggest moon of Saturn, where average temperatures hover around −180° Celsius (−292° Fahrenheit). Maybe Titan is populated by large *very* slow-moving creatures that look more like living rocks than bug-eyed monsters. The original series of *Star Trek* had a great episode called "The Devil in the Dark" with aliens like this.

There are even more radical ideas out there. In a famous science-fiction novel, *The Black Cloud*, astronomer Fred Hoyle imagined an interstellar cloud of gas that was not only alive but also conscious. Electrical currents running through the cloud formed the basis of its nervous system. This idea is certainly fun to consider, but it's very hard to understand how evolution could ever have led to such a sentient cloud.

Put it all together, and the consideration of non-carbon-based life forces us to be humble. Nature is way more creative than humans. It's likely to be more clever than we are about how and where it makes life. On the other hand, we don't want to be too humble. The rules of chemistry are the universe's, not ours. Those rules, embodied in that periodic table of elements we all slept through in high school, will be king. Alien life may definitely be strange, but it might not be stranger than we can imagine.

TALKING TUMBLEWEEDS OR FLYING FORESTS
What will aliens be like?

We've seen that chemistry tells us the best bet is for aliens to be carbon-based, like us. But what will they look like? Will they have legs? Will they have wings? Will they have pointy ears and antennae poking from their foreheads? Will they even have heads to "fore" with at all? Or will they look like mobile shrubbery or something else even weirder? To answer these questions, we move from physics and chemistry over to the elegant logic of biological evolution.

It's been said that all biology would be meaningless without the theory of evolution pioneered by Charles Darwin back in 1859 (though he began formulating it in 1838). Darwinian evolution can be summed up in two simple terms. The first is *descent with modification*. Living systems reproduce themselves by passing their characteristics down to their progeny. Baby birds get wings. Baby fish get gills. That's the "descent" part. Sometimes, however, the copying involved in reproduction gets messed up. A baby crow, for example, might be born with a much longer or shorter beak than its parents. That's the "modification" part.

Next comes something you've no doubt heard of: *survival of the fittest*. This simply means that, as environments change, the organisms that are best adapted to those changes are the ones that go on to reproduce. If longer beaks are more useful in the new environment, then long-beaked crows thrive. After a few generations, you'll see only long-beaked crows. The shorter-beaked crows get outcompeted generation by generation until finally they are gone.

That's evolution in a nutshell. But will it apply to aliens?

It's hard to imagine anything we'd reasonably call life that does not reproduce or respond to its environment. Those two characteristics are part of almost every definition of living organisms anyone has ever come up with. But that means it's also hard to imagine any form of life that doesn't involve Darwinian evolution. If you are better adapted to the environment, you reproduce more. If you reproduce more, your line continues, and you win the game of life. Put it all together, and it does seem that the logic of Darwinian evolution must apply to life as a universal phenomenon. Anywhere life appears in the universe, it will be shaped by descent with modification and survival of the fittest. In other words, Darwin rules aliens.

Now we can try combining the universal laws of evolution with the universal laws of physics and the universal laws of chemistry. We'll do this by starting with something we can all agree on: stuff (i.e., matter).

Alien planets will be made out of stuff. They have to be made of stuff because everything in the universe is made of stuff. Matter, and energy flows between pieces of matter, are the basic building blocks of the physical universe. From the laws of physics and chemistry, we know that stuff can come in only a few basic forms: solids, liquids, and gases. Let's also make the reasonable assumption that staying alive always involves finding energy to stay alive—by eating other stuff. This gives us a basic problem we can think about in our attempt to conceive of the forms alien bodies might take. From a physical, chemical, and evolutionary perspective, the options for how aliens find meals depends on whether they live in a solid, a liquid, or a gas.

We'll investigate gases first.

Say you want to move around on a planet that's mostly gas, like a super-Earth with a huge atmosphere. The first thing you'll need is some way to counter gravity so you don't sink downward to the dark, nasty, high-pressure planetary depths. The physics of gases allows only certain ways to accomplish this. You could float. You could push jets of gas downward to maintain your position. You could also find a way to get air flowing over a curved surface. That last one turns out to be familiar, so we'll focus on it.

When air flows over a downwardly curved surface, the pressure on the bottom side is higher than on the top. That pressure imbalance pushes the surface up, providing *lift* against gravity. This may sound abstract, but what I am talking about here is a *wing*. Wings are one solution to the problem of moving around in an atmosphere. On Earth, the process of evolution has solved that problem with wings a lot. Insects have wings. Birds have wings. Bats have wings. In fact, wings evolved in separate evolutionary lineages four different times over the last five hundred million years. Over and over again, evolution exploited the physics of gases by discovering curved surfaces for flying.

The independent emergence of wings multiple times is an example of what scientists call *convergent* evolution, and it's incredibly important for thinking about aliens. Convergent evolution happens because the physical world imposes constraints on life that will lead it to find the same solution to the same problem whenever that problem arises. There are multiple examples of this principle during Earth's multi-billion-year history. We've already discussed wings, but there are also the various kinds of eyes, which use biologically produced lenses, and the streamlined body plans of the dolphin (a mammal) and the shark (a fish).

Convergent evolution tells us that we shouldn't be surprised to find life with wings on planets with atmospheres. That's pretty cool. However, the convergence principle does *not* tell us what those wings will look like or how biology will structure them. Maybe they aren't made from feathers or skin but look more like soap films in the form of a thick mucus layer drawn across a fiber frame (I just made that up). The details will be different on different worlds, but as long as fluid flows can generate lift, some kind of wing will be on a planet's evolutionary menu.

Taking this physical, chemical, and evolutionary approach is, however, not so straightforward. You can't just declare, "Wings for everyone!" and be done with it. Astrobiologists must also think long and hard about how planets and their physics can corral the choices. Consider, for example, a super-Earth that is similar to our world but way more massive. Naively we might expect that kind of planet to be inhabited by thick, sturdy creatures. They'd be built to

DO ALIENS DO IT TOO?

withstand high gravity, like mega-elephants thunderously plodding across the landscape. However, evolution might use the planet's high gravity in a completely different way. An atmosphere on a massive Earth-like planet would be much denser than ours as its atoms get pulled more strongly toward the surface. The physics of gases tells us it's much easier to generate lift in a dense atmosphere. You'd hardly need to flap your wings at all. It would be more like swimming than flying. So life on a super-Earth might forgo the ground almost entirely. Evolution would exploit the ease-of-flying conditions to create an entire biosphere of creatures that touched down only once or twice in their whole life, perhaps to breed.*

Convergent evolution also can't tell us which trajectory evolution might pick for a given set of planetary conditions. On a gaseous world, will we end up with winged creatures, balloon-like organisms, or animals that hover using bio-jets? What about solutions that haven't been used on Earth at all? The physics of interfaces between different physical phases is fascinating. To move along the interface between a solid and liquid, like the bottom of the sea, or a solid and a gas, like the land and the atmosphere, some version of a leg seems like a good idea. Evolution on Earth has solved the problem of moving across interfaces via movable, jointed sticks (legs) many times. Although the number of legs ranges from two (humans) to hundreds (millipedes), they are, in the end, all legs. Score one for convergent evolution.

What about wheels? There is no evolutionary lineage on Earth that has chosen wheels as its method for navigating interfaces. Why is that? Maybe it's impossible to create a wheel on an axle via biological molecules. Maybe even if you could create the wheel-axle combination, it would be impossible to create a drive belt to power it. There doesn't seem to be a physics or chemistry reason

---

* This was the premise of episode 2 in the great Netflix documentary *Alien Worlds*. You should watch it. I got to participate on the episode about civilizations, and they took me to the gates of Area 51. No one shot at me.

167

why evolution can't solve this problem with wheels, but on Earth it has chosen not to.

All this tells us that, while convergent evolution is a powerful idea that we can apply to aliens, we should *always* expect nature to be way more inventive than we are. When life is confronted with a planet hosting conditions different from Earth's, we can expect Darwinian evolution to exploit physics to solve its basic problems. But we can also expect that we probably aren't smart enough to know exactly how the problem gets solved.

There is, however, another side of evolution that works in the opposite direction; it goes by the name *contingency*. Basically, what we're talking about here are accidents. Evolution works via random mutations. In this way, it's "blind." Evolution doesn't have a plan. Instead, it relies on two different kinds of accidents occurring over and over again.

The first kind of accident is the random mutations that show up in random offspring. An example of this would be some random Klingon baby somewhere born with a random mutation in the genetic code that manifests as a spiny ridge on her forehead. The second kind of accident is the fit between mutations and the environment. If the mutation happens to make a Klingon kid fitter to survive in a changing environment, then the kid and the mutation go on. Maybe the planet Kronos (for non-nerds that's the Klingon homeworld) suddenly gets shrouded in dark clouds from an asteroid impact, and our baby girl's forehead ridge is full of neurons that allow her extra sight in the dark. Now she's fitter for survival.

Of course, if the mutation does not make the kid fitter, that may be the end of the line for the mutation. Because the next accident builds on the last one, the specific details of a species—like its body shape—are contingent. All the accidents must happen in exactly the sequence they did, or things would have worked out differently.

And we're talking about a lot of accidents here. Evolution works on timescales of millions, tens of millions, or even hundreds of millions of years. Because most individual organisms on Earth live fewer than a hundred years, there are a mind-numbing number of generations, accidental mutations, and accidental environmental changes for evolution to run through.

Contingency was a primary concern for the great evolutionary biologist Stephen Jay Gould. In his book *Wonderful Life: The Burgess Shale and the Nature of History*, Gould considered all the fossilized creatures that are no longer around today and yet represent the ancestors of all the life that *is* around. His conclusion was that, if Earth's history were rewound and started over, everything would have turned out differently. Other accidents—a rock tumbling off a cliff at just the right moment a billion years ago, a lightning strike killing a mutated dinosaur at another moment a hundred million years ago—would have led evolution in entirely different directions. On that alternative Earth, there would be no bears or whales or bees or humans. There would still be a lot of life, but it wouldn't be the life we're familiar with.

For evolutionary biologist Simon Conway Morris, however, there are only a handful of specific kinds of evolutionary pressures that the environment can supply to organisms (gravity, resistance, heat, cold), and these always lead evolution in the same direction. Morris takes the opposite position from Gould and has gone so far as to claim that the human body plan will be repeated on other planets. We should, he says, expect to meet creatures that look like us. It's worth noting that Morris is most definitely in the minority among biologists on this count. What is particularly cool about this whole debate, however, is that recently a way may have appeared to end it with experiments.

Since 1988, a team of biologists at the University of Michigan has been tracking the evolution of generation after generation of a species of bacteria called *Escherichia coli*.[1] They fill beakers with these little bugs and keep them fed and happy. The microbes then reproduce, creating one generation after the next. Every so often, the team removes some of the critters and freezes them. This basically preserves a snapshot of their evolutionary trajectory. The unfrozen bacteria populations continue living their bacterial lives. They keep reproducing and, in the processes, keep evolving new characteristics. The University of Michigan scientists have, for example, seen the bacteria evolving toward much bigger cell sizes.

Eventually distinct populations emerge. There's a line of small-cell bacteria and a different line of large-cell ones. Once this evolutionary divergence occurs, the researchers can unfreeze the ancient (by the bacteria's standards) ancestors and run the tape of life again. If the ancestors eventually produce offspring with the same evolutionary divergence, it would be a win for convergent evolution. If they don't, then contingency is king. After watching more than sixty thousand generations of *Escherichia coli* bacteria evolve, the weight of evidence so far seems to favor contingency. Accidents matter more than convergence in the specifics of how creatures evolve.

If this is true, then we are indeed the only human beings in the universe. On no other world would we expect all the zillions of accidents from microbes to simple multicellular creatures to large multiorgan animals to play out in the exact same way. On no other world would evolution lead to alien humans or even humanoids.

This battle between convergence and contingency tells us we'll need to be prepared for surprises. That, finally, takes us back to alien sex.

Sex is a great idea. I don't mean for you personally but for all life in general. By mixing the genetic information from different individuals, life vastly increases it ability to produce winning combinations of traits that can endure in different circumstances. Will evolution on other worlds also discover sex rather than just having creatures clone themselves forever? Convergent evolution would argue that it's too good a strategy *not* to emerge on other worlds as it did on Earth. However, that doesn't mean that a future Captain Kirk or Captain Janeway will be able to get it on with an alien counterpart.

Biologically, sex is defined entirely as the swapping of genes between different individuals. There are so many ways this might happen that the principles of convergent evolution might not be much help. Bacteria have a kind of sex that involves what's called horizontal gene transfer, where microbes spray their surroundings with the genes that they "want" others to take on.* That doesn't sound particularly romantic—unless you're a bacterium, I guess. So while sex makes sense for aliens, the forms it will take are anyone's guess.

ALIEN MINDS
Can you talk with an ET?

We are the cosmic lonely hearts. Ever since *Homo sapiens* first evolved about three hundred thousand years ago, we've had only one another to talk with. Lord knows, you can't have much of a conversation with a plant (OK, some folks say they can). And while some animals seem to talk with one another, they can't, or won't, talk with us. I think I know what my dog is thinking, but despite my best efforts, he won't say a word to me.

If only we had some other intelligent species to compare notes with. What could someone (or something) who didn't share our evolutionary and historical baggage teach us? Maybe they could give us their take on the big questions of philosophy. Maybe they could tell us if they believed in God or not. Anything they could say about what it's like to *not be us* would be revolutionary. This dream of interstellar, interspecies communication was at the heart of many early efforts in SETI, and it still fuels much of the enthusiasm for UFOs. Unfortunately, it's also a dream that ignores a central question about aliens.

Will their minds be anything like our own?

Asking about alien minds takes us deep into questions about minds in general. We all agree that humans are self-conscious beings, but

---

* It's actually more complicated than "spraying." Genes from dead bacteria can be picked up by others, or the genes can be transferred directly by linking cells up, or they can be sent from one to the other.

we're not entirely clear how self-conscious other animals are. And what exactly is consciousness anyway (with or without a self)? For some scientists and philosophers, we are basically computers made from meat. For them, our brains have the proper neural circuitry for generating the *feeling* of being conscious, and that's the end of the story. According to this view, if I knew all the connections in your brain and could re-create them in a computer, I would basically re-create *you* in a computer. For other smart people who think about this, however, that idea is nothing but wishful thinking. In their view, minds are a lot more than just neural hardware and software. Instead, a mind is always "embodied." Minds don't just ride around in our heads but are more like ecosystems that require the living experience of *being in a body* that is itself part of a wider community of living things.

These two perspectives are pretty different, and they lead to two very different views of alien minds.

If brains are nothing more than computing machines, then all that matters are the mathematical rules for computing. As long as those rules are the same for us as they are for an alien species that evolved on an exoplanet, then we should be able to understand each other. We would both be living in the same kind of world, shaped by the universal laws of physics that are themselves expressed in the universal rules of mathematics. If that's the case, questions about alien minds really become questions about alien mathematics. Are the mathematical rules humans use the same ones all aliens use? In other words, is mathematics universal?

Carl Sagan, among others, was hopeful that this was the case. Sagan always believed that if we encountered an alien species, we could learn to communicate with it by starting with basic mathematical relations: $1 + 1 = 2$, $2 + 2 = 4$, and so on. Eventually, each species would decipher the other's way of saying more complicated things. Here "complicated" means something like "the circumference of a circle is two times its radius times pi." Because pi equals 3.14159 . . . for us, it must equal 3.14159 . . . for them too. Once we ironed all that out together, we'd be on our way to deciphering the rest of each other's language. It would just be a matter of time, then, until we were sharing knock-knock jokes and swapping cures for cancer.

However, a story like this works only if mathematics really is universal. It demands that math be something our minds *discover*, something already true about the world. For many scientists and philosophers, though, mathematics isn't something we discover. It is something we invent. We make it up, but it's still incredibly useful. This way of thinking about mathematics often sees its success at describing the world as rooted in our experience of living in our kind of body. It's a language, just like English or Spanish or Chinese. Like all our languages, mathematics is something uniquely human, and that's why this view has huge implications for understanding aliens. Creatures that aren't human and don't have our kind of bodies would have a different experience. That would lead to different minds and different mathematics.

Imagine creatures that don't have hands or fingers. Perhaps they're intelligent protoplasmic blobs, like giant amoebas. If they needed to grab something, they'd extend a tendril from their blobby bodies and then retract it. Would these aliens even have a concept of discrete numbers? Would they have integers like 1, 2, 3, etc.? Maybe they wouldn't "see" the world as discrete objects to be counted, as we do with our two hands with five fingers each. Their math might not resemble anything we are used to or could even understand.

I'll give you an example that someone else already thought up. One of my favorite alien contact movies, *Arrival*, explored this possibility by taking it to an extreme. After giant seven-legged aliens arrive in giant floating spaceships, humans send a physicist and a linguist to establish communications. The physicist, a Carl Sagan–like character, tries using mathematics to communicate. He immediately and spectacularly fails. The linguist, who understands how languages arise from embodied experience, breaks through, but only after realizing that the aliens experience time in an entirely different way than we do. The heptapods, as they are called, experience the past, present, and future at once. They have embodied a physics that's entirely different, and that difference is reflected in their language. It's an exhilarating idea. Maybe the real "book of physics" is much bigger than what we humans experience. Maybe alien species get access to different chapters of that book. That would mean that we and the aliens were, to some degree, living in different worlds.

There is one more possibility, which is kind of terrifying, so I left it for last because I didn't want to scare you. (You're welcome.) Maybe the consciousness of aliens is no consciousness at all. Most of our science-fiction stories teach us to expect aliens who are self-reflecting, self-aware creatures like us. We might be experiencing different versions of the universe, but at least we're both *experiencing* something. But what if there are aliens that are intelligent but not conscious? Imagine, for example, that an alien race creates machines that rapidly outpace their creators. The machines become essentially alive. They are highly intelligent and can solve the problems of reproduction and finding energy rapidly and effectively. They spread throughout the universe in powerful spaceships, chewing up one star system after another for their needs. Through it all, however, they are never self-aware. They do not have the capacity to reflect on what they are doing because there is no "light" inside. There is no interior dimension to their minds. In philosophical terms, we can ask what it's like to be a bat or a horse or a dolphin, but we couldn't ask what it's like to be them. It's not like anything, because they don't experience anything at all.

There is no need for these kinds of aliens to be machines. Perhaps biological evolution can produce such intelligent but non-conscious life. We could *never* talk to creatures like these. There simply wouldn't be anyone *in there* to talk to. If they were intent on swallowing the Earth for its resources, there would be no one to plead with. They'd just be empty shells completing tasks.

The possibilities for alien minds span this wide spectrum I've just laid out. There are the hopeful and familiar versions of alien minds, with whom we hang out at the *Star Wars* cantina and tell each other stories of our exploits on Altair IV. There's also the strange and fascinating version in which we may never fully understand each other but we recognize our identities as fellow beings in the vast cosmos. Finally, there's the "Run!" version, which is simply about not getting eaten by uncaring unconscious entities. It's the sweep of these possibilities—all the ways aliens could be so deeply different—that will make *contact*, if it happens, the most revolutionary event in all human history.

ALIEN ETHICS
Should we hide or fire a flare?

When the alien spacecraft first landed, there was panic. Soon, however, it was clear that they were here to help. The aliens showed us how to cure cancer and end war. Their aim, they said, was for us to thrive. All they wanted was "to serve man," which was even the title of a book they carried everywhere. When the aliens offered us the chance to visit their home world, thousands of people took them up on it. One of their vast spacecraft was readied for the trip. Then, just as the door of the packed spaceship was closing on the human passengers, news came that some of the alien book *To Serve Man* had finally been translated beyond the title.

As one of his friends is stepping into the doorway, the translator runs up screaming, "It's a cookbook!"

This version of contact comes from a classic short story by Damon Knight, which was turned into the best *Twilight Zone* episode ever. What makes the plot so effective is how it neatly reveals the poles in our thinking about alien ethics. On the one hand, many people assume that any species technologically advanced enough to travel between stars must also be ethically advanced. War and predation must be horrors they left in their deep evolutionary past, unlike us. They might even be so evolved and so advanced to appear as angelic beings of pure good and light. On the other hand, other people assume "as below, so above." Why shouldn't the worst traits in our species be the ones that win out on a cosmic scale, as they often do on Earth? Perhaps there are nothing but wolves out there among the stars.

Trying to answer this question using anything more than storytelling is hard. While we might be able to use the principles of convergent Darwinian evolution to infer something about the biology of aliens, understanding what ethics guide their behavior takes us even deeper into questions of alien minds and alien society than we've gone before.

One clue from life as we know it, however, is the balance between species that evolve from predators as opposed to those that begin as prey. On Earth, many of the species that show the highest degrees of intelligence are hunters. This makes sense. The evolutionary battle for survival will favor hunters who best anticipate their prey's future behavior. A lion that consistently misses the direction of the gazelle's leap won't be queen of the jungle very long. Nevertheless, intelligence as a hunter is not enough. Evolving into a technological civilization also requires cooperation on a grand scale. That fact may mean that only *social* predator species go on to travel the stellar byways.

The fact that predators might be the evolutionary template for star-traveling species sounds like bad news for us in terms of becoming somebody's lunch. The record on Earth, however, gets complicated when we look at the species most like humans. On the one hand, there are the savanna chimpanzees (*Pan troglodytes*), who are a violent bunch. Chimp tribes routinely go to war with each other. They kill members of other tribes (including their babies), and within a tribe, the alpha male chimp tends to rule by intimidation and the threat of violence. Then there are the bonobos. These are the forest chimpanzees (*Pan paniscus*), who are evolutionarily as close to us as the savanna chimps. Bonobos are matriarchal and they are far more peaceful. They tend to use sex as a way of calming disputes, and they almost never engage in violence against each other. I don't know about you, but I'd definitely rather hang in the trees with the bonobos!

In considering these differences between our evolutionary cousins, we must also consider that chimps evolved where food was very scarce. In their original natural environments, they were, as an anthropologist friend put it, hungry and miserable most of the time. On the other hand, bonobos emerged in parts of Africa that had plenty of food. Their lives were relatively easy. This difference may explain a lot about how their social behavior ended up and may prefigure what happens with alien civilizations. Whether we meet free-loving hippie bonobos or terror-happy fascist chimpanzees may hinge on the details of the environments their planets offered.

Thinking about alien ethics takes us in another direction, however, that centers on our own behavior. In 1974 Frank Drake pointed the giant Arecibo radio telescope at M13, a cluster of stars at the edge of the Milky Way. Then he beamed a message designed to announce our existence to the galaxy. The message, if aliens could decode it, gives a vague idea of where we humans live and even provides some details about what we look like. It was both revolutionary and controversial. The version of SETI that Drake originally created meant just passively listening for aliens. When he took it to the next level, this new endeavor of messaging extraterrestrial intelligence (or METI) meant actively letting them know that we're here. Since Drake's first transmission, there have been others who've purposely beamed high-intensity transmissions toward specific cosmic locations. The ethics of their activity has, however, been sharply criticized.

The anti-METI critique takes two forms. The first tack might be termed "Who speaks for Earth?" Given the uncertainties surrounding alien civilizations, why should a few astronomers with access to radio telescopes get to decide when and how we announce ourselves? Shouldn't the process be decided by the global community as a whole? The second criticism focuses solely on the consequences of METI. The physicist Steven Hawking was explicit in his warning that our own history shows that contact between civilizations with different levels of technology doesn't usually work out well for the one with less aggressive tech. It's probably much better to keep our heads down. This is the moral of the wonderful science-fiction series that began with *The Three-Body Problem* by Liu Cixin. In those books, Liu explores the dark-forest hypothesis, which says the reason we haven't observed any signals from alien civilizations yet is that they're all hiding. Like creatures in a dark forest, they know better than to start squawking and give away their positions.

There is a counter-argument to the METI-haters, which says that our radar and TV transmissions have been announcing our presence to the universe for many years anyway. However, the counter-counter to this position is that, in reality, those signals are pretty weak. If aliens didn't already know where to look, they'd probably miss us. In the end, we are left with the all-important question of who is eating whom. Should we be hiding, or should we expect that anyone who has made it to the stars will become an agent of peace? If they're like us, we should probably shut up, but whichever choice we make, it better not be the wrong one.

WILL THE BIOLOGICAL ERA BE SHORT?
Welcoming the robot overlords

Goo.

When you come right down to it, that's what life on Earth is. Sure, it might be encased in rough bark or a hard tortoise shell, but somewhere in there, even if you have to go down to the cellular level, life is goo. As a result, when we imagine alien life, there's a tendency to emphasize the goo part—goo dripping off fangs (the *Alien* franchise), goo that takes over and uses human hosts (Marvel's *Venom* and Robert Heinlein's novel *The Puppet Masters*), and goo that digests humans (*The Blob*). Lots and lots of goo.

DO ALIENS DO IT TOO?

What if most aliens aren't gooey? That is, what if intelligent life's biological phase is relatively short? Maybe every planetary alpha species eventually leaves those kinds of bodies behind by uploading their minds into silicon-based computers and living forever as digital entities.

The duration of the biological era for intelligent life is a question that's become more pressing for astrobiologists with the rapid rise of digital technologies. Our first electronic computer, called ENIAC (electronic numerical integrator and computer), was built in 1945 to handle calculations involved in nuclear-weapons design. Its circuits were made from copper wires and glass vacuum tubes, and they filled an entire room. Today your cell phone has millions of times more computing power than ENIAC did. That exponential rise in processing speed has been matched by an equally astonishing increase in software capacities. Advances in artificial intelligence (AI) have already reshaped the world and with new advances like ChatGPT things are going to get even more reshape-y. AI runs the autonomous robots that sort your purchases in Amazon warehouses, and it's AI that lets us have conversations with a virtual global machine named Siri.

This rapid advance of digital machines in the form of networked computers and robots is what fuels the *transhumanist movement* and its hopes for the Singularity. For transhumanists like Ray Kurzweil, as computers become more sophisticated, there must come a time when a machine will match our mental capacity for designing the next, better generation of machine. Once that machine-designed machine is built, it's game over. From that moment on, computers will be better at designing better computers than humans. The consequence is a kind of runaway machine evolution with digital "intelligence" quickly outpacing that of humans. This runaway is the transhuman Singularity. On the other side of the Singularity, computers will become so sophisticated that we humans should be able to merge with them. At least that's the claim. All of us will be able to download ourselves into silicon form. We will live inside of computers as virtual versions of ourselves. Moreover, because

179

machines can be repaired or updated, our digital versions will be functionally immortal. We will live forever.

The transhumanists' vision can be either intoxicating or horrifying, depending on your inclinations. Regardless of your stand on their aspirations, though, the basic idea that intelligent species who develop technology might eventually outgrow their need for bodies does seem at least possible. If it *is* possible, won't that be the route most long-lived alien species will take? Biological life is always an ongoing struggle to maintain bodies against the onslaught of time. Because that struggle is always doomed, what civilization *wouldn't* take the option for another, more enduring form as machines? Even if some species decide not to download, you could argue that evolution would eventually weed them out on galactic or cosmic scales. The downloaded, machine-melded versions of life would be far smarter and stronger than their goo-bound cousins. Eventually only the silicon versions of life would remain.

There's another possibility that's perhaps even more frightening (again depending on your inclinations). Maybe machines don't want our stinking downloads. From *The Terminator* to *The Matrix* to *Westworld*, science fiction is full of robot-rebellion stories. Once a biologically based species creates artificial intelligence, those sentient machines may simply decide they've had enough. At best, they might leave their creators (that's us) to their fate and head off to the stars. At worst, the machines might decide extinction was a proper option for their inferior gooey progenitors. Even without hostile intentions, self-replicating machines could be the engines of galactic settlement. Von Neumann replicators (named after mathematician John von Neumann) would be purely AI-driven starships (no goo aboard) that would arrive in a system, strip it for resources to build more starships, and leave.

Whether it's ancient downloaded intelligences, evil robot exterminators, or blind Von Neumann replicators, the arguments we've sketched out above raise the radical possibility that our search for life won't find life at all. Our gooey biological bodies may already be

DO ALIENS DO IT TOO?

obsolete on the cosmic marketplace of possibilities. If this bums you out, then you'll be happy to hear that there are counter-arguments to the inevitable rise of the robot overlords.

First of all, it's not clear that AI will ever achieve anything like the kind of "general" intelligence with consciousness that we possess. On Earth right now, we see amazing advances in AI in the form of "machine learning." These are the kinds of AI-driven programs that can recognize faces, play chess, fabricate images, and even have limited (sometimes scary) conversations with humans. But the thing about all these versions of AI is that they are brittle. They are trained to do specific things. Ask them to do something different and they can fail miserably. Supporters say it's just a matter of time until we have an AI that can do anything you ask, just as a human can (i.e., an AI with a general, not a specific, kind of intelligence). That may be true, but they've been saying that for seventy years. Even the latest "generative AI" systems, like the different versions of ChatGPT, are nowhere close to having our kinds of minds. For all their power, these systems are still just (very powerful) prediction machines. They use vast amounts of data that's been fed into them to make statistically based predictions about a best response to a query, kind of like auto-complete on steroids times a billion. Despite all their power, they don't really *know* anything, because there is no one in there to do the knowing. That means it is possible that our kind of general intelligence simply can't be captured in silicon form.

An even more potent criticism is that the transhumanist vision of downloading into a computer may also not be possible. Underlying that transhumanist dream (or fantasy) is the philosophical assumption that humans are basically just meat computers. According to transhumanists, everything that makes you what you are—all your internal experiences of love and dread and hope and compassion— is just circuitry. But there are lots of scientists and philosophers who think this idea is very, very wrong. While being conscious certainly involves having the neural circuitry of a brain, it also involves a whole lot more. This "more" includes being in a body that's part of both natural and social ecosystems. From this point of view, the whole "download yourself into a computer" thing is deeply clueless about the meaning of *yourself*.

Finally, there is also the possibility, suggested by astrobiologist Caleb Scharf, that the silicon switch won't be permanent. Scharf emphasizes that biological evolution is very good at solving problems, so even if a species went the machine route, they might eventually opt to go back to biology again after a few billion years. Think of all the zillions of amazing chemical calculations happening in your cells right now, keeping you alive. In the end, biology may just be better at certain kinds of things. If that is the case, then machine life would be like adolescence. It would just be a phase that most forms of interstellar intelligence pass through (hopefully without the mood swings). As you can see, these questions are pretty open-ended right now.

There are good arguments to expect alien robot overlords, and there are also good reasons for intelligence to stay gooey. In the end, as with so many other things about aliens, we won't know until we look, and we *are* finally looking.

ANCIENT ALIENS
How to think about million-year-old civilizations

What happens to really, really old civilizations? This is one of the most difficult and important questions facing astrobiology and technosignatures science. We may not be able to search effectively for intelligent life in the universe if we don't know how really old intelligences behave. We won't know what to look for. It's also a question that bears directly on our own fate. What happens if we get past climate change, if we get past the threats of nuclear war, if we figure out how to build AIs without having them kill us all or turn us into batteries? What do our long-long-term possibilities look like? What might we, or any civilization, become if given enough time?

DO ALIENS DO IT TOO?

To set the stage, let's remember that humanity as a technological civilization is only about a hundred years old. I mean, we've been leaving technosignatures that aliens could detect—our radio signals, city lights, and atmospheric pollution—for only about a century. Of course, we've had agriculture going for at least nine thousand years, and we were hunter-gatherers for long before that. Each of these forms of social organization were complex and rich and varied. However, ancient Rome or Tang Dynasty China are not the kind of civilization we're talking about when we talk about SETI and technosignatures. We have just a century of being a radio technological species, not very long compared to the age of the universe. Our relative youth has important consequences in the search for aliens.

Back in 2020, David Kipping, Caleb Scharf, and I worked out the probabilities surrounding the age of any alien civilizations we might detect. To be fair, it was really Kipping, a very smart theoretical astrophysicist, who did the mathematical kung fu. Scharf and I cheered him on, then checked the calculations and then added some additional ideas. Here's the question we wanted to answer: Should we expect to discover alien civilizations that are younger, older, or the same age as ours? The answer, using fairly sophisticated probabilistic reasoning, turned out to be older. Probably even *much* older.

If our calculations were right, when we do find aliens, they'll have been in the technological civilization game for a long, long time. While we've been fooling around at building jet planes, electrically powered cities, radio transmitters, and even crude spaceships for a century, they will have been civilization-ing for millions or even billions of years. This conclusion is interesting by itself, but really it raised a much bigger, much freakier question.

What are million- or billion-year-old civilizations like? What happens to them over that unimaginable stretch of time?

The longest civilizations on Earth, like the Egyptian or Chinese dynasties, lasted only a few thousand years. A ten-million-year-old alien civilization would be about ten thousand times older than our longest-lived societies. What possible forms of organization can intelligent creatures create after being around for that much time? How do we even think about civilizations over such mind-melting stretches of history? Just as important, does *anyone* make it that long?

To be specific, let's formulate a specific question. Would a ten-million-year-old civilization be a single continuous entity, or would it rise and fall many times during those eons?

Even more compelling than the continuity of the civilization is the continuity of the civilization builders. How would they change over ten million years? If they engage in a goo-to-silicon switch, does that mean that the society, its ambitions, and its actions fundamentally change? Do they give up on expanding, or do they conquer the galaxy in the name of computers? Even if they don't switch to machine form, ten million years is long enough for Darwinian evolution to significantly alter the biology of our aliens. Ten million years ago we were small, hairy orangutan-like creatures. Now look at us! Clearly a lot can happen in ten million years.

Beyond natural evolution, could we expect our aliens to employ some version of genetic engineering to sculpt their bodies and minds across the eons? We probably will do that over just the next few centuries, so who knows what form aliens might engineer themselves into. Perhaps a ten-million-year-old civilization turns itself into a planetwide forest in which every "tree" is a kind of sentient biological supercomputer networked to every other tree. The planet itself would then become a nearly immortal living, thinking being.

The planetary forest example raises another possibility. As a civilization advances and its technological capacities increase, a point might be reached where it disappears into the laws of physics themselves. The idea that a truly ancient and advanced race might eventually become pure energy makes its appearance a lot in science fiction (*Star Trek* has used the plot point a lot). But here we're going even further to imagine the aliens weaving themselves into the fabric of reality itself. Check out *Interstellar* for an example of this kind of idea, a story in which our far-future selves manipulated space-time to allow our current selves to stave off extinction. Cool and confusing at the same time.

Thinking about truly ancient civilizations is both exhilarating and frustrating. On the one hand, the possibilities are so expansive, it feels like staring over a cliff stretching down for miles. Vertigo is definitely an appropriate response when trying to imagine a million years of evolution. On the other hand, it can feel almost impossible to say anything about ancient civilizations that's not just a story. There are so many unknowns, and almost none of them have the kinds of constraints science can supply to guide our thinking. Technology, for example, might make our ancient aliens into gods, or it might stall out at some far lower level. For one example, maybe rapid interstellar travel is just not possible.

Despite the difficulties, thinking about the long-term evolution of civilizations is an essential part of technosignature studies. We just can't avoid it. If we're looking for alien civilizations and those alien civilizations will be much older than us, we'll have to figure out how to find ancient alien technosignatures too.

Finally, there's another reason why we need to think about the long-term evolution of alien civilizations. We want to become one. We need to figure out how to endure. We can hope that other species have faced down the perils of nuclear war or climate change or who knows what else lies ahead of us. We can hope that others have gotten past these challenges. In that way, the astrobiological work of mapping out the possible long-term histories of alien civilizations also becomes an exercise in imagining our own possible future. Maybe no one makes it more than a few hundred years. That would suck. Or maybe half of the species that get to our phase go on for another million years. That would be better.

Which option is it? Who makes it and for how long? What are the attributes of a species that *does* build a long-lived technological civilization? Answering those few questions might be the best reason of all to search for cosmic life.

CHAPTER 8

Why Aliens Matter

IT'S MORE THAN YOU THINK

Back in the late 1970s, when I was sixteen, I'd take the train across the Hudson River from New Jersey into Manhattan every chance I could. New York was a glorious mess back in those days, kind of like a postapocalyptic amusement park. The pee stench in the subways that were scrawled with graffiti. The streets so trash strewn, they made their own gravy when it rained, as David Letterman once joked. For me, it was all a giant adventure, from trying to slip into clubs like CBGB or Kenny's Castaways to the mostly illegal, twenty-four-hour festival that was Washington Square Park. You should know this by now, but I'm more into science than even the best party (not that anyone would invite me anyway), so the high point of those trips was always the Hayden Planetarium. This was back in the pre-digital day when the planetarium was all hand-painted dioramas of the Martian surface and mechanical scales that told your weight on Jupiter (450 pounds!). I spent endless hours in the Hayden getting high on star shows and big glossy images of interstellar clouds. I'm not gonna lie; I once cried with joy in front of Albert Einstein's statue.

After every one of those visits, I'd emerge back onto West Eighty-First Street with my mind reeling. Heading into Central Park, I'd wander for hours, letting the noise and the bustle bring me back to the same blur of questions over and over again.

All those stars. All those possibilities. Is this happening anywhere else? Are there other cities on other worlds filled with others like me going about their day?

Looking back over all the decades, it's stunning how much has changed when it comes to those questions. I hope this book has given you a sense of the gulf between then and now. Back then, the Hayden had no exhibits about the exoplanet census, because no one knew if even one exoplanet existed. There was no diorama with a super-Earth landscape, because no one had imagined there might be super-Earths. You also wouldn't find a big section at the Hayden dedicated to finding biosignatures or technosignatures, because those terms hadn't even been invented.

So, yeah, everything has changed. The search for aliens is a real thing now, and science has gone all in on it.

All that progress still leaves another question hanging, though. Why does it matter? What will change if we do find alien life? Will it be a revolution in human understanding, as when we realized that the Earth wasn't flat or that we weren't the center of the universe? Or will it be big news for a week, and then everyone'll go back to their social media feeds for daily doses of political outrage and dance videos?

Some people answer this question by saying it depends on what we find. The discovery of bacteria on Mars won't matter as much as finding that UAPs really are alien visitors, they say. But after all we've just learned, I want to consider what finding even the simplest forms of life would mean for us.

The weirdest thing about being human in the early twenty-first century is that we know so much and yet we're still so completely clueless. When it comes to the basic question about what we are and what we're supposed to be doing, all our learning just highlights how deep we are in the dark. For the first time in history, we know how vast the universe is and how rich it is with planets and their possibilities. Yet we still don't know if we—and by "we," I mean all life on planet Earth—are a one-off. We still don't know if we are a cosmic accident that happened only once and only here, and then never happened again.

Here's the thing. If we had just *one* other example of life forming, if we had just *one* other instance where a bunch of nonliving chemicals got transformed into a living organism, then we'd know we're not a crazy accident. We would know that *it*, meaning life, had happened twice. And if it could happen twice, it could happen three times—or thirty or thirty thousand or thirty billion.

Having just one other example of life blows open the door of cosmic possibilities because life is *different*. Life is unlike any other physical system in the entire universe because life *creates*. Tell me everything there is to know about some star out there right now, and I can tell you what its entire future will look like. A star will never surprise me, because its behavior (and its future) is set by the laws of physics. Life is certainly governed by the laws of physics, but there's more. Life is a complex system wherein the laws of biology (i.e., evolution) also apply. That means life surprises. Life invents. Life goes beyond itself.

We've seen how Earth's life hijacked the planet, how its ability to spread and invent changed the entire history of this world. Based on this, I can't help but wonder how far life's inventive powers might go in terms of changing the universe. We saw via the Fermi paradox that even one spacefaring species could settle the entire galaxy in a short time. By considering the Kardashev scale, we saw how type 3 civilizations could harvest the energy of entire galaxies. When we asked about really old civilizations, we wondered if they weren't able to alter the laws of physics. Maybe our universe itself is an offshoot of an experiment or game some super-advanced species is messing with. Maybe life, as it evolves and creates and invents, changes the universe as much as it's already changed the Earth.

We don't know—we *can't* know—what life is capable of. As Ian Malcolm (Jeff Goldblum's character) said in the first *Jurassic Park* movie, "Life finds a way." On a cosmic scale we can now begin to ask, "Find a way to do what?" The answer could be "anything and everything." If we find even one other example of life out there, the "space" of what can happen instantly explodes. Everything, meaning the whole universe and our place in it, instantly gets way more interesting.

So, the detection of a biosignature would be a game changer. Finding evidence of an alien civilization, however, would take us to a whole other level. There are a couple of ways to see why that kind of discovery would matter. First, there is the simple fact that such a detection shows us that what's happened here, in terms of evolution from microbes to a complex technological species, happened somewhere else. A bucket of questions would spill out from such a discovery, and each one has the power to reshape what we know about ourselves. Are the creatures who built that alien civilization social in the same way we are? How do they organize themselves? Are they a hive mind, or are they collectives made of individuals like us? Do they use symbols and process information in the same way we do? What kinds of technologies do they use? Can we recognize the principles those technologies are built on? Can we learn enough about their technology from a distance to teach us anything that could be useful here and now?

Most of all, though, the important thing about discovering an exoplanet civilization is that it would be just the beginning. If we found a planet with technosignatures, the race would be on to build bigger, better, and more advanced telescopes. In the years that followed, we would—pardon my language—study the shit out of that civilization-bearing world.

We would also have to decide about making contact. Should we begin a very slow conversation? Remember, if they were fifty lightyears away, every call-and-response would be a hundred-year cycle. On the other hand, maybe we'd think better of dropping a dime (for all our young readers, that means making a call). Maybe we'd decide that just watching for a while was the smart play. Eventually, we might even start sending robot probes across the stars to give us a better look, and in the process, make those probes into UFOs to the aliens.

Without a doubt, humanity would never be the same after discovering an alien civilization. It doesn't mean we'd necessarily start acting different, as in being nicer to each other. (I wish.) But in the longer term, questions raised by such a discovery could change our religions (did Jesus come to save the aliens too?), our ethics (are our cherished values universal or just provincial?), and our art (how do we represent ourselves in books, movies, and songs knowing there are others so different?).

There's another really important way the discovery of other civilizations would matter. It could show us our own future. That's the other weird thing about being human in the early twenty-first century. It's not clear there's going to be a twenty-second.

Over the last hundred years, we've developed extraordinary, powerful technologies that reshaped the planet and society. Now many of those technologies also seem poised to kill us. Whether it's climate change (our energy-harvesting tech), artificial intelligence (our information-processing tech), or good old nuclear war (our military tech), it's not entirely clear how much longer all this can last. I don't think any of those threats will make humans extinct. Extinction is a tall order. But each holds the possibility of spinning things out of control enough to take down the global technological civilization we all depend on for, you know, staying alive. That's where those aliens come in.

Finding even one civilization further along than we are would show us it's possible to get there. As of right now, we don't know if the universe actually "does" high-tech, long-term sustainable civilizations. We know the universe makes stars. We know it makes comets and black holes. We've seen lots of those things. But it's entirely possible that no civilization anywhere in the cosmos gets much past where we are before flaming out. Finding another civilization would be what mathematicians call an existence proof. It would show us that such a thing—the kind of civilization we need to become— can exist. That would be a big relief. It would give us a lot of hope. Also, by studying that world and that civilization, we might get clues about what it takes to become a long-term, sustainable high-tech civilization. We might understand something about how a planet and a society work together to allow both the biosphere and the technosphere to thrive.

Finally, if UFOs turned out to be alien spacecraft, well, then . . . um . . . well . . . *holy bleeping crap!* You don't need me to tell you why that discovery would matter. In one fell swoop it would mean: (a) life exists on other worlds; (b) civilizations exist on other worlds; and (c) they are here, and they are going to eat, kill, teach, or mate with us.

There is a big addendum to that last "holy crap," though. Imagine that we can conclusively prove that UFOs move in ways that no human technology could produce, but that's it. What if that's all we get? They still don't land in Paris and climb out of their spaceships. They still don't sit still long enough for us to get a good look at them. In that case, we'd have to just keep doing science on them as we'd do with any other difficult-to-observe phenomena. Would it then be a good idea to try to capture one? Should we try to shoot one down? Even if we had proof that they were alien in some way, we might never make contact. That wouldn't diminish the importance of the discovery, though. Even without a direct human-to-alien contact, we'd have proved that something truly not human was prowling our skies. That knowledge would take us down the road to more questions and more science about those questions. That's more or less how any scenario associated with aliens will work anyway. Each question leads to the next and the one beyond it. That's what makes it all so exciting.

Finding aliens would be the greatest scientific achievement of all time, and like all truly great scientific discoveries, it would change the trajectory of human history. That is why the progress we've made toward answering this ancient mystery matters. That is why the fact that we're standing on this shoreline, readying our boats to be pushed into the sea to begin this grand journey, matters. You, me, and every other person alive today—we're the lucky ones. We all carry the questions our ancestors asked about life and the universe, but we alone get to be there when the answers emerge.

Enough waiting. Enough talk. The time has come to find out for ourselves.

Let's go!

Acknowledgments

This book was so much fun to write, and I am so very grateful to all the people who helped me along the way. Before I tell you about them and how awesome they are, I first have to tell you that while I worked hard to make sure I got all the details right, any errors that slipped past me are completely my fault.

Now on to the gratitude. First and foremost, I have to thank my friend who also happens to be my agent, Howard Yoon. The idea for this book was a collaboration between the two of us. I knew I wanted to write something that let people see how much was changing in the search for cosmic life, but Howard saw how to broaden its scope. The chapters on UFO tech were his idea, and they were a gas to think about and write. Thank you, Howard.

Next, I need to let you know about Sarah Haugen, my editor at HarperCollins. Sarah's insight took the idea for the book to the next level while giving me the space to explore different directions during the writing. Her edits were sharp, to the point, and vital. They allowed me to write in the voice I'm most comfortable with while keeping the narrative from wandering off into scientific or personal tide pools. I am deeply grateful for her support and her work. Thank you, Sarah.

I had a number of readers who looked at early versions of the book and provided feedback. I'm very thankful for their time and insights. Morgan Ryan started out as my fact-checker and then became a kind of guardian angel in terms of both facts and style. Our conversations wandered from protein structure to Led

Zeppelin minutia, and they were all enjoyable. Dani Buchheister, a whip-smart Penn State astrobiology graduate student was also a great help. I asked the wonderful author Sarah Scoles to look over the chapters about UFOs and UAPs, and her expertise was greatly appreciated. I am thankful to Jonathan Sherwood, a talented science-fiction writer, for his insights. I am also very grateful to the Quane Brothers for their support throughout the project, including my *The Division 2* partner-'n-rampage Jack Baumgartner for an early read. My good friend Thad Spencer not only read the manuscript but was invaluable in thinking about the cover art. Those discussions took us into some great territory about the purposes of design and illustration.

I need to give a very special set of acknowledgments to the members of our NASA Categorizing Atmospheric Technosignatures (CATs) grant team. Many of them show up in one way or another in the book. Our work together these past few years has been a joy and very inspiring. The team includes my co-investigators on the grant Jacob Haqq-Misra, Ravi Kopparapu, Manasvi Lingam, Sofia Sheikh, and Jason Wright along with the super-talented students Macy Huston, Nick Tusay, and Connor Martini. I sought advice from Jacob and Ravi many times and they were kind enough to read early versions of chapters. Jason is my go-to font of knowledge for everything SETI both in terms of history and current science.

At the University of Rochester I also need to thank Peter Iglinski and Lindsey Valich for all their help getting the word out there on astrobiology. I am also grateful to Michael Osadciw for his wonderful design ideas for the book cover.

Finally I need to thank Woody Sullivan, Dan Watson, Eric Blackman, Jonathan Carrol-Nellenback, Gavin Schmidt, David Kipping, Caleb Scharf, Gourab Ghoshul, Marcelo Gleiser, Evan Thompson, John Dunne, Joan Halifax, Amedeo Balbi, Sarah Walker, David Grinspoon, and Damian Sokwinski for the most excellent conversations. I also

ACKNOWLEDGMENTS

thank my lifelong friends Robert Pincus and Paul Green who are helpfully skeptical of my rantings and will always argue with me about anything or just make me laugh.

As we all know from the *Fast and Furious* franchise, in the end it's all about family. So Harrison, Sadie and Myani, Elisabeth and Hendrick, Leon . . . Thank you. I'd be lost without you.

And then, always and forever, my beloved Alana.

Notes

Chapter 1: How Did We Get Here?
How Our Ancient Questions About Aliens Took Their Modern Form

1. An account of the discussion between Fermi and his friends can be found in a paper by Eric M. Jones, "'Where Is Everybody': An Account of Fermi's Question," (Los Alamos National Laboratory, Mar. 1985), https://www.osti.gov/biblio/5746675.

2. Megan Garber, "The Man Who Introduced the World to Flying Saucers," *Atlantic*, June 15, 2014, https://www.theatlantic.com/technology/archive/2014/06/the-man-who-introduced-the-world-to-flying-saucers/372732/.

3. Russell Lee, "1947: Year of the Flying Saucer," National Air and Space Museum, Smithsonian Institution, June 24, 2022, https://airandspace.si.edu/stories/editorial/1947-year-flying-saucer.

4. Donovan Webster, "In 1947, A High-Altitude Balloon Crash Landed in Roswell. The Aliens Never Left," *Smithsonian Magazine*, July 5, 2017, https://www.smithsonianmag.com/smithsonian-institution/in-1947-high-altitude-balloon-crash-landed-roswell-aliens-never-left-1809 63917.

5. Kal K. Korff, "What Really Happened at Roswell," *Skeptical Inquirer*, vol. 21, no. 4, July–Aug. 1997, https://skepticalinquirer.org/1997/07/what-really-happened-at-roswell.

6. Edward J. Ruppelt, *The Report on Unidentified Flying Objects* (New York: Doubleday, 1956). See chapter 2. Also see: "Estimate of the Situation," especially note 9, https://military-history.fandom.com/wiki/Estimate_of_the_Situation#cite_note-9.

7. Sarah Scoles, *They Are Already Here: UFO Culture and Why We See Saucers* (New York: Pegasus Books, 2020), 69.

8. *Report of Meeting of Scientific Advisory Panel on Unidentified Flying Objects Convened by Office of Scientific Intelligence*, CIA, Jan. 14–18, 1953, The Black Vault, https://documents.theblackvault.com/documents/ufos/robertsonpanelreport.pdf.

9. Scoles, *They Are Already Here*, 74.

10. Scoles, *They Are Already Here*, 78; and "Project Blue Book—Unidentified Flying Objects," National Archives, https://www.archives.gov/research/military/air-force/ufos.

Chapter 2: So How Do We Do This?
Critical Ideas That Shaped, and Still Shape, Our Search for Aliens

1. Keith Cowing, "Independent Review of the Community Report from the Biosignature Standards of Evidence Workshop," press release, National Academies of Science, Oct. 3, 2022, https://astrobiology.com/2022/10/independent-review-of-the-community-report-from-the-bio signature-standards-of-evidence-workshop.html.

2. Su-Shu Huang, "The Problem of Life in the Universe and the Mode of Star Formation," *Publications of the Astronomical Society of the Pacific* 71, no. 422 (Oct. 1959): 421–24.

3. Hannah Ritchie and Max Roser, "Energy Production and Consumption," Our World in Data, https://ourworldindata.org/energy-produc tion-consumption.

4. Jason T. Wright, "Dyson Spheres," *Serbian Astronomical Journal* 200 (2020): 1–18, https://arxiv.org/abs/2006.16734.

5. Jonathan H. Jiang et al., "Avoiding the Great Filter: Predicting the Timeline for Humanity to Reach Kardashev Type I Civilization," *Galaxies* 10, no. 3 (May 2022): 68, https://arxiv.org/pdf/2204.07070.pdf.

NOTES

Chapter 3: WTF UFOs and UAPs?
How They Do, or Do Not, Fit into the Search for Aliens

1. Stephen J. Garber, "Searching for Good Science: The Cancellation of NASA's SETI Program," *Journal of the British Interplanetary Society* 52 (1999): 3–12, https://history.nasa.gov/garber.pdf; and Daniel Oberhaus, "A Brief History of Scientists Searching for Extraterrestrial Life," *Motherboard*, Vice, Dec. 4, 2015, https://www.vice.com/en/article/jmaawd/a-brief-history-of-scientists-searching-for-extraterrestrial-life-124.

2. The story of the hoax has been covered in many places, like "How an Alien Autopsy Hoax Captured the World's Imagination for a Decade," *Time*, June 24, 2016, https://time.com/4376871/alien-autopsy-hoax-history/; and Neil Morris, "The Alien Autopsy . . . Oh, No, Not Again!" Personal Pages, University of Manchester.

3. "UFO Encounter 1: Sample Case Selected by the UFO Subcommittee of the AIAA," *Astronautics and Aeronautics*, July 1971, http://kirkmcd.princeton.edu/JEMcDonald/mcdonald_aa_9_7_66_71.pdf.

4. James E. McDonald, "Science in Default: Twenty-Two Years in Inadequate UFO Investigations" (General Symposium, Unidentified Flying Objects, American Association for the Advancement of Science, 134th Meeting, Dec. 27, 1969), http://kirkmcd.princeton.edu/JEMcDonald/mcdonald_aaas_69.pdf.

5. Helene Cooper, Ralph Blumenthal, and Leslie Kean, "Glowing Auras and 'Black Money': The Pentagon's Mysterious U.F.O. Program," *New York Times*, Dec. 16, 2017, https://www.nytimes.com/2017/12/16/us/politics/pentagon-program-ufo-harry-reid.html.

6. ABC News, "Pentagon Declassifies Navy 'UFO' Videos," April 27, 2020, https://www.youtube.com/watch?v=lWLZgnmRDs4; https://www.youtube.com/watch?v=2TumprpOwHY.

7. Bill Whitaker, "UFOs Regularly Spotted in Restricted U.S. Airspace, Report on the Phenomena Due Next Month," *60*

Minutes, CBS News, https://www.cbsnews.com/news/ufo-military-intelligence-60-min utes-2021-05-16.

8. Keith Kloor, "The Media Loves This UFO Expert Who Says He Worked for an Obscure Pentagon Program: Did He?" *Intercept*, https://theinter cept.com/2019/06/01/ufo-unidentified-history-channel-luis-elizondo-pentagon.

9. Scoles, *They Are Already Here.*

10. Bryan Bender, "Ex-Official Who Revealed UFO Project Accuses Pentagon of 'Disinformation' Campaign," *Politico*, May 2, 2021, https:// www.politico.com/news/2021/05/26/ufo-whistleblower-ig-complaint-pentagon-491098.

11. Mick West, "I Study UFOs, and I Don't Believe the Alien Hype: Here's Why," *Guardian*, https://www.theguardian.com/comment isfree/2021/jun/11/i-study-ufos-and-i-dont-believe-the-alien-hype-heres-why.

12. *Preliminary Assessment: Unidentified Aerial Phenomena*, Office of the Director of National Intelligence, June 25, 2021, https://www.dni.gov/files/ODNI/documents/assessments/Prelimary-Assessment-UAP-20210625.pdf.

13. Keith Kloor, "Pentagon UFO Study Led by Researcher Who Believes in the Supernatural," *Science*, June 29, 2022, https://www.science.org/content/article/pentagon-ufo-study-led-researcher-who-believes-supernatural.

Chapter 4: What If They Are Aliens? If UFOs Are ET, How'd They Get Here, and What the Hell Are They Doing?

1. Some of these ideas first appeared in my column "If UFOs Are Alien Spaceships, How Did They Get Here?" Big Think, Nov. 11, 2021, https://bigthink.com/13-8/ufo-light-speed.

2. David G. Messerschmitt, Philip Lubin, and Ian Morrison, "Optimal Mass and Speed for Interstellar Flyby with Directed-Energy Propulsion," June 27, 2022, arXiv, https://arxiv.org/abs/2206.13929.

NOTES

3. Dennis Overbye, "Reaching for the Stars, Across 4.24 Light-Years: A Visionary Project Aims for Alpha Centauri, a Star 4.37 Light-Years Away," *New York Times*, Apr. 12, 2016.

Chapter 5: Cosmic Curb Appeal?
Where to Look for Aliens

1. Comets, Vanderbilt University, https://www.vanderbilt.edu/AnS/physics/astrocourses/ast201/comets.html; and Daisy Dobrijevic and Charles Q. Choi, "Comets: Everything You Need to Know About the 'Dirty Snowballs' of Space," last updated Jan. 18, 2023, Space.com, https://www.space.com/comets.html.

Chapter 6: The Cosmic Stakeout: How We're Going to Spy on ET

1. Now much of the focus is on the TESS mission, which took over after Kepler completed its run. TESS stands for Transiting Exoplanet Survey Satellite. See: "Planet Hunters TESS," Zooniverse, https://www.zooniverse.org/projects/nora-dot-eisner/planet-hunters-tess.

2. Thomas G. Beatty, "The Detectability of Nightside City Lights on Exoplanets," *Monthly Notices of the Royal Astronomical Society* 000 (preprint, Feb. 21, 2022): 1–12, https://arxiv.org/pdf/2105.09990.pdf.

3. Svetlana V. Berdyugina and Jeff R. Kuhn 2019, "Surface Imaging of Proxima b and Other Exoplanets: Albedo Maps, Biosignatures, and Technosignatures," *Astronomical Journal* 158 (Nov. 25, 2019): 246.

4. Manasvi Lingam and Abraham Loeb, "Natural and Artificial Spectral Edges in Exoplanets," *Monthly Notices of the Royal Astronomical Society: Letters* 470, no. 1 (Sept. 2017): L82–L86, https://doi.org/10.1093/mnrasl/slx084.

Chapter 7: Do Aliens Do It Too?
What Will We Find When We Find Aliens?

1. Richard E. Lenski, "Convergence and Divergence in a Long-Term Experiment with Bacteria," *American Naturalist* 190, no. S1 (Aug. 2017), https://www.journals.uchicago.edu/doi/full/10.1086/691209.

Recommended Reading

Good for general audiences

Phillip Ball, *The Book of Minds*
Martin Beech, *Terraforming*
Lee Billings, *Five Billion Years of Solitude*
A. Allsuch Boardman, *An Illustrated History of UFOs*
Milan M. Ćirković, *The Great Silence: Science and Philosophy of Fermi's Paradox*
Paul Davis, *The Eerie Silence*
Steven J. Dick, *Astrobiology, Discovery and Societal Impact*
Steven J. Dick, *Plurality of Worlds: The Extraterrestrial Debate from Democritus to Kant*
Donald Goldsmith, *Exoplanets*
David Grinspoon, *Earth in Human Hands*
Kevin Hand, *Alien Oceans*
Arik Kershenbaum, *The Zoologist's Guide to the Galaxy*
Kal K. Korff, *The Rosewell UFO Crash*
Simon Conway Morris, *From Extraterrestrials to Animal Minds*
Kart T. Pflock, *Rosewell*
Mark Pilkington, *Mirage Men*
Dirk Schulze-Makuch and William Bains, *The Cosmic Zoo*
Sarah Scoles, *They Are Already Here: UFO Culture and Why We See Saucers*
Peter Ward and Donald Brownlee, *Rare Earth*
Steven Webb, *Where Is Everybody: 75 Solutions to the Fermi Paradox*

More advanced books

Manasvi Lingham and Avi Loeb, *Life in the Cosmos*
Douglas A. Vakoch and Maureen Dowd, *The Drake Equation: Estimating the Prevalence of Extraterrestrial Life Through the Ages*

Index

abiogenesis (creation of life from nonlife)
    Drake's equation and complexity of civilizations, 16, 19
    Miller-Urey experiment and, 103–105
Advanced Aerospace Threat Identification Program (AATIP), of Pentagon, 69–71
Air Force plane crews, sightings of UFOs, 64–65
Alcubierre, Miguel, 84
Alien Autopsy (FOX television documentary), 63–64
alien life, impacts of finding of, 186–192
alien life, search for, generally
    early days of, xiii–xiv, xv*n*
    Fermi's paradox and Great Silence, 11–12
    heliocentric versus geocentric conceptions of space, 4–6
    modern advances in, xv–xviii
    negative effects of UFO reports on, 58–61
alien life, search methods after exoplanet discoveries, 127–158
    alien megastructures, 139, 140–143
    artifacts in solar system, 148–150
    artificial lights and, 145–147
    beacons and, 138
    biosignatures, 48–49, 129–134
    'Oumuamua and, 150–154
    possible characteristics of life, 159–185
    pollution and, 143–145
    technosignatures, 136, 137–140
    technospheres, 134–136
    terraforming, 154–157
alien life, steps toward science of, 35–55
    ancient thinking and, 1–6
    Arnold and UFOs, 19–23
    Drake and Project Ozma, 38–43
    Dyson and megastructures, 47–50
    Fermi's paradox and, 6–12, 44, 188
    government misinformation and, 26–31
    habitable zones and, 43–47
    Kardashev and advanced civilizations, 51–55
    Roswell and spread of

INDEX

misinformation, 23–26
science fiction and popular culture and, 31–34
standards of evidence and, 36–38
alien megastructures
   Dyson and, 44, 47–50, 51, 53
   search for aliens and, 139, 140–143
alien optimists, 3, 5, 12–13
alien pessimists, 2–3, 12–13
alien technology, and physics of universes, 78–98
   inconsistencies of, 86–90, 96–98, 150
   interstellar travel's challenges, 79–86
   lack of scientific data about, 98
   multidimensions and, 90–96
Alien Worlds (documentary), 167n
Alpha Centuri, 81, 129
American Association for the Advancement of Science, 29–30, 65
amino acids
   found in interstellar clouds, 104–105
   molecular life and, 102
Ancient Aliens, 72
anomaly searches, for evidence of aliens, 139, 142–143, 148–150
Anthropocene, 136
Aristotle, 2–3
Arnold, Kenneth, 20–24, 26
Arnold, Luc, 142
Arrival (film), 173–174
artifacts of other civilizations, in solar system, 148–150

artificial intelligence (AI) anomaly searches and, 139, 143, 149
   digital advances and, 179–180, 190
artificial lights, search for technological civilizations and, 145
astrobiology, xiv
astro-engineering, 50
astronomy, in 1950s, 40–41
atmospheric characterization, biosignature detection and, 131–133, 135, 143–145
atomists, 3
Aztec, NM, UFO hoax and, 62–63

Baum, L. Frank, 41
beacons, SETI and search for, 59, 138
Beatty, Thomas, 146
Behind the Flying Saucers (Scully), 62
Bialy, Shmuel, 153
Bigelow, Robert, 69
biosignatures
   search for life and, xv, 129–134, 138
   standards of evidence and, 133–134
biotechnical probability, of life on exoplanets, 124–126
Black Cloud, The (Hoyle), 163
Boyajian, Tabatha, 141–143, 153
Breakthrough Listen initiative, 149
Breakthrough Starshot, 83
brine, origin of life and, 107–108
Bruno, Giordano, 4–5
Bryan, Richard, 60

Calvin, Melvin, 13
carbon, origins of life and, 160–163
Case of the Ancient Astronauts, The (PBS documentary), 57
Cassini, 107
Catholic Church, ancient questions of extraterrestrial life and, 3, 4–5
Chariots of the Gods (von Danikien), 57
ChatGPT, 179, 181
chemical equilibrium, life and biospheres, 132
Chicago Sun, 21
Chlorof luorocarbons (CFCs), Earth's technosignature and, 144–145, 157
Citizen Science project, 141
civilizations, use of term, 7n.
    See also technological civilizations
Clarke, Arthur C., 150
climate change, xvii, 18, 54–55, 77, 115–116, 135, 144, 190
Cold War, UFO sightings and, 20, 27, 30–32, 51–52
comets, and origin of water on planets, 119, 155
Condon, Edward, 29–30
Condon Committee Report, 29–30, 57, 94
    McDonald and critiques of, 65–66
Conte, Silvio, 60
contingent evolution, 168–169, 170
convergent evolution, 166–168, 170
Conversations on the Plurality of Worlds (De Fontenelle), 5
Copernicus, Nicolaus, 4

cryosleep, possible interstellar travel and, 82, 85

Darwin, Charles, 5, 164–165
Day the Earth Stood Still, The (film), 33
De Fontenelle, Bernard, 5
deoxyribonucleic acid (DNA), molecular life and, 102, 104
Dick, Steven J., 2n
Dietrich, Alex Anne, 68–69
dimensions
    hovering and antigravity technology and, 88–90
    humans and three-dimensions, 91
    mathematics and hyperdimensional possibilities of aliens, 92–94
Drake, Frank, 52, 58, 60
    beacons and, 138
    METI and, 177
    Project Ozma and, 38–43, 51
    see also Drake equation
Drake equation, 44, 59
    exoplanets and odds of other technological civilizations, 122–126
    in formula form, 14
    habitable zones and, 7, 12–19
    scientific questions raised by, 19
    in sentence form, 14–15
    technological civilizations and, 12–14
    terms of, explicated, 15–18
    use of, 18–19
dwarf stars, habitable planets around, 110–111

INDEX

Dyson, Freeman, 44, 47–50, 51, 53
Dyson rings, 50
Dyson spheres, 49–50, 142
Dyson swarms, 50, 142

Earth
    abiogenesis and, 120
    astronomy and concept of rarity of, 5
    atmosphere of, 157
    biosignature of, 129, 132
    as Kardashev type 0.7 civilization, 53–54
    snowball phases of, 115–116
    study of and insights on exoplanetary evolution, xiv as terrestrial world, 112
    techno signature of, 144–145, 157
    water and, 118
ecumenopolis (city-world), 146
Einstein, Albert. *See* general theory of relativity (GR)
Einstein-Rosen bridges (wormholes), 83
electromagnetic radiation (waves), search for aliens and, 39, 59–60, 87, 88, 130, 136
Elizondo, Luis, 69–70
Enceladus (moon of Saturn), 107–108
energy, and foundations of scientific search for extraterrestrial life, 47–50, 52
entropy, thermodynamics and, 54
Epicurus, 3, 126
Epsilon Eridani, 40
Escherichia coli, evolution and, 169–170
Estimate of the Situation (US government), 27–28, 73
ethics, of aliens, 175–178
Europa (moon of Jupiter), 106–107
evolution, 164–165
    contingent, 168–169, 170
    convergent, 166–168, 170
    Drake's equation and, 15–19
    predators, prey, and ethics, 176–177
existence proof, impacts of finding of alien biosignatures, 190–191
exoplanets (extrasolar planets), xiv
    Drake equation and odds of other technological civilizations, 122–126
    impacts of finding of, 111, 127, 129
    search for, 99–100, 109–111
exotic matter, 83, 84
Extremely Large Telescope, 158

Fermi, Enrico, paradox of, 6–12, 44, 188
    importance of, 11
    SETI and Great Silence, 11–12
51 Pegasi b (exoplanet), 109–110
flourine, 160
"flying saucers," Arnold's sightings and lessons of, 20–23
Forbidden Planet (film), 33
force bosons, gravity and, 88–89
force particles, gravity and, 88–89
forward-looking infrared (FLIR) cameras, videos of UAPs and, 67–68

207

frameshift drives, possible interstellar travel and, 84
Fravor, David, 68

Galileo Project, 77
Garber, Steven, 61
gas giant planets, 112, 113
GeBauer, Leo A., 62
general theory of relativity (GR), interstellar travel and space time, 83, 84–85, 94, 95
generation ships (century ships), possible interstellar travel in, 82
geocentric model of space, 4–6
GIMBAL videos, of UAPs, 68
GOFAST videos, of UAPs, 68, 73–74
Gould, Stephen Jay, 169
graviton, 88–90
gravity
    antigravity technology and, 83, 88–89
    evolution and, 165–166
    nongravitational acceleration and, 152
    ocean moons and, 107
"great oxygenation," 132
Greek astronomers and philosophers, questions of extraterrestrial life and, 2–4
Green Bank Observatory, 13, 39–41
Greenewald, John Jr., 70
Grusch, David, 72–73

Habitable Worlds Observatory, planned, 43, 43n, 158
habitable zones
    Drake's equation and, 7, 12–19
    and foundations of scientific search for extraterrestrial life, 43–47
Haldane, J.B.S., 102–103
Haqq-Misra, Jacob, 66, 145
Hart, Michael H., 9
Hawking, Steven, 177–178
Hayden Planetarium, 186–187
heliocentric model of space, 4–6
high-beam argument, for observations about UFOs, 97, 150
hoaxes, regarding UFOs, 61–62
    Aztec, NM and, 62–63
    Roswell, NM and, 63–64
Holocene climate state, 116
horizontal gene transfer, 170–171, 171n
Hottel, Guy, memo of, 63
Hoyle, Fred, 163
Huang, Su-Shu, 45
hydrogen, absorption of light and, 130
Hynek, J. Allen, 94–95
hyperdrives, possible interstellar travel and, 84

ice ages, as planetary events, 115
ice giant planets, 112–113, 114
inertia, UFO travel and, 89–92
intelligence
    Drake's equation and complexity of civilizations, 16–17
    and possible alien life lacking in consciousness, 172, 174–174
Intercept, The, 70
interdimensional hypothesis, of Vallée, 94–95
Interstellar (film), 117, 184–185

INDEX

interstellar settlements, simulation of, 9–10
interstellar travel
    alien technology and constraints on, 85–86
    possible means of, 82–85
    speed of light and, 80–81
    vast distances between stars and, 79–80
Invasion of the Body Snatchers (film), 33
James Webb Space Telescope (JWST), 43, 143, 145, 158
Jet Propulsion Laboratory, 54, 149
Jung, Carl, 94
Jupiter
    distance from Earth, 108
    exoplanets as "hot" Jupiter's, 110, 131
    as gas giant, 112, 120n
    ocean moons of, 100, 106–107, 120n
    size and gravitational pull of, 107, 186
    visible from Earth, 2

Kardashev, Nikolai, 44, 51–55, 188
Kepler Space Telescope, 110, 140–141, 142
KIC 8462852, megastructures and, 141–143
Kipping, David, 183
Kloor, Keith, 70
Kopparapu, Ravi, 66, 145
Kurzweil, Ray, 179

life
    abiogenesis and, 16, 19, 100–105

carbon and origins of, 160–163
    changes to planets made by, 99–126, 132–136
    as creator and inventor, 188–189
    exoplanets, Drake equations, and odds of other technological civilizations, 122–126
    exoplanet habitable zones and, 108–111
    hallmarks of, 101–102, 105, 164–165, 168–169
    ocean moons and, 106–108
    as originating on planets, 109
    oxygen and, xiv, 132–133
    "post-biological" machine-based, 82
    snowball worlds and, 115–121
    super earths and, 112–114
light
    biosignatures and absorption of, 130
    hospitable zones and, 46
    interstellar travel measured in light-years, 80–81
light sails, possible interstellar travel and, 82–83, 153–154
Lilly, John, 13
Lingam, Manasvi, 147
Liu Cixin, 178
Loeb, Avi, 77, 147, 153
Lovelock, James, 131–132, 133
Lubin, Philip, 82–83
"lurkers" (hypothetical probes), 149–150

Mars
    atmosphere of, 157
    climate state of, 100, 116

NASA probes of, 42, 132
size of orbit of, 109
terraforming and, 154–155, 156
as terrestrial world, 112
visible from Earth, 2
water and, 118
McDonald, James E., 30, 64, 65–67
Mercury
lack of atmosphere, 154, 157
size of orbit of, 109
temperature on, 16, 46
as terrestrial world, 112
visible from Earth, 2
water and, 118
messaging extraterrestrial intelligence (METI), 177–178
metabolism, as hallmark of life, 101–102
metamaterials, UFOs and, 87
microbes, life and, 128, 128*n*, 134, 156, 169–171, 189
Milky Way galaxy
Drake and questions of advanced civilizations, 14, 18–19, 177
Fermi's paradox and, 8–9
size, by light year, 81
Miller, Stanley, 103
Miller-Urey experiment, abiogenesis and, 103–105
Milner, Yuri, 83, 149
minds, of aliens, 171–175
intelligent but non-conscious life and, 172, 174–175
mathematics and computing, 172–174
molecular life, components of, 102
Moon, NASA artifacts on, 148–149

Morris, Simon Conway, 169
Musk, Elon, 154
Muslim societies, questions of extraterrestrial life and, 4
mutation, as hallmark of life, 101–102, 105, 168–169

NASA
early SETI efforts of, 59–61
grants for study of exoplanet technosignatures, xvii
Habitable Worlds Observatory, planned, 43, 43*n*, 158
images of Europa, 106
Mars probes of, 132
Planet Hunters website, 141
planetary missions of, 42–43
Roswell and UFO panel of, 26
technosphere research of, xvii, 74, 77, 134, 137–140, 145
National Academy of Science (NAS), 13, 43, 59
Navy pilots, UFO videos and, 26, 58, 67–68, 69, 71, 73–74
Neptune, as gas giant, 112–113
New Horizons space probe, 80–81
New York Times, 67–68, 69
Newton, Isaac, 5
Newton, Silas M., 62
non-carbon based life, universal chemical properties and unlikelihood of, 160–163
non-conscious life, aliens and, 172, 174–175
noosphere, 135. *See also* technospheres

ocean moons and worlds, 106–108,

INDEX

118–121, 120n
octopus, brain of, 120
Oort Cloud, 81
Oparin, Alexander, 102–103
'Oumuamua, 150–154
oxygen
  biosignature detection and, xiv, 132–133
  red-dwarf planets and, 133–134

Panoramic Survey Telescope and Rapid Response System (PAN-STARRS), 151
Pearman, J. Peter, 13
periodic table, chemistry of life and, 161–162
photons, gravity and, 88–89
physics. *See* alien technology, and physics of universes
physiology and behavior, expectations for aliens, 159–185
  biological era, transhuman movement, and criticisms of, 178–182
  chemistry and non-carbon-based life, 160–163
  ethics and, 175–178
  importance of thinking about long-term evolution and, 182–185
  minds and, 171–175
  physical forms and convergent versus contingent evolution, 164–171
planets, origins of life on, 109
Plurality of Worlds (Dick), 2n
Pluto, 80
pollution, search for

technosignatures and, 144
polymers, molecular life and, 102
"post-biological" machine-based life, 82
Project Blue Book, 29, 66–67
Project Cyclops, 59
Project Grudge, 28
Project Mogul, 26, 30
Project Ozma, 38–43, 44, 51, 138
Project Saucer, 27
Project Sign, 27–28
proteins, molecular life and, 102, 104
Proxima Centauri, 129
Proxmire, William, 59
Puthoff, Harold E., 70

quantum mechanics, possible interstellar travel and, 84–85

radio telescopes, 39–43
  Drake and, 13–14, 17, 39–40, 177
  exoplanet discovery and anomaly searches, 138–139
  Project Cyclops and, 59
  Project Ozma and early, 39–43
  SETI and Great Silence, 11
red-dwarf planets, 111, 133–134
Reid, Harry, 69
Rendezvous with Rama (Clarke), 150
reproduction, as hallmark of life, 101–102, 105, 164–165
ribonucleic acid (RNA), molecular life and, 104
ribosomes, molecular life and, 102
Robertson Panel (CIA), 28

Rochester Museum and Science Center, 115
Roswell, NM
   government misinformation and myths about, 26–31
   negative effects on scientific research, 23
   UFO hoax and, 63–64
Roswell Report: Case Closed, The (government report), 26
runaway effect, on snowball planets, 117
Ruppert, Edward, 28

Sagan, Carl, 13, 58
   Earth and Kardashev scale and, 53
   expansion of alien species and, 9, 10
   mantra regarding evidence, 57
   mathematics and communication with aliens, 172–173
   Proxmire and SETI, 59
Santilli, Ray, 63
Saturn
   distance from Earth, 108
   as gas giant, 106, 107, 112
   ocean moons and, 100
   visible from Earth, 2
savanna chimpanzees, evolution and, 176–177
Scharf, Caleb, 182, 183
science
   as conservative, 36, 40
   constrained imagination and, 45
   standards of evidence and, 36–38, 133–134

Science, 72
science fiction, UFOs and popular culture, 31–34
"Science in Default" (McDonald), 66
Scoles, Sarah, 27, 27n, 70
Scully, Frank, 62
search for extraterrestrial intelligence (SETI). *See alien life entries*
second law of thermodynamics, 48–49, 54
sensitivity, as hallmark of life, 101–102
sex, aliens and, 170–171
Sheikh, Sofia, 139
silicon, possibilities of noncarbon-based life and, 161–163
Singularity (transhuman movement), 178–180
60 Minutes, 68
snowball planets, 115–118
sodium, absorption of light and, 130
solar collectors, search for technological civilizations and, 147
solar sails, possible interstellar travel and, 82–83
solar system SETI, 148–150
solar systems, architecture of all, 109
Soviet Union, 51–52
space, as three dimensional, 91
space-time, 53, 83–85, 94
spiritualist movement, fourth dimension and, 94
standards of evidence
   biosignatures and, 133–134
   modern science and, 36–38

INDEX

Stapleton, Olaf, 48
Star Maker (Stapleton), 48
Star Trek (television programs), 87, 89, 163, 184
stars
    Drake's equation and complexity of civilizations, 15
    lack of life on, 99
string theory, 96
Struve, Otto, 13, 39–40
Sullivan, Woody, 122–126
super earth enigma, exoplanets and, 112–114

Tarter, Jill, 12, 58, 60, 137–138, 157
Tau Ceti, 40
technological civilizations
    Drake and, 14
    Fermi's paradox and, 7–12, 7n
    importance of thinking about million-year-old, 182–185
    Kardashev's levels of, 52–53
    see also alien life, search methods after exoplanet discoveries
technosignatures
    Kardashev and, 52
    search for alien life and, xv, xvn, 136, 137–140, 142
    "Technosignatures Workshop" of NASA, xvii, 74, 77, 134, 137–140, 145
technospheres, 134–136
temperature, hospitable zones and, 45–46
terraforming, signs of, 154–157
terrestrial worlds, 112, 118
thermal vents
    life on Earth and, 107, 120
    ocean worlds and, 121
thermodynamics, second law of, 48–49, 54
They Are Already Here: UFO Culture and Why We See Saucers (Scoles), 27n, 70
Three-Body Problem, The (Liu Cixin), 178
time
    abiogenesis and, 105
    differs from other dimensions as being only one way, 95
    space-time, 53, 83–85, 94
Titan (moon of Saturn), 163
To Serve Man (Knight), 175
transhuman movement, 178–180
transit method, exoplanet detection and alien megastructures and, 140–141, 142
    appearance of transits, 141
    biosignatures and, 100, 129–131
Twilight Zone (television program), 175

unidentified aerial phenomena (UAP), xv–xvi
    becomes government's official name for UFOs, 57, 57n
    invisibility poorly done by supposed, 87, 150
    military pilot videos and government programs about, 67–74
    need for actual evidence, scientific data, and rational research about, 73, 74–77
    Roswell and, 26

unidentified identified flying objects (UFOs), xv, 56–77
   Arnold and press misinformation about, 19–23, 24
   as artifacts of environment, 65, 98
   connections with alien life, 56, 75
   Fermi's paradox and, 8, 10
   hoaxes and, 61–64
   impacts of finding of alien biosignatures and, 191
   McDonald and critiques of, 64–67
   negative effects on SETI, 58–61, 137
   physics and behavior of, 38, 87
   popular culture and, 31–34, 56
   science and standards of evidence required to study, 57–58, 98
University of California–Santa Barbara, 82–83
University of Colorado–Boulder, 29
University of Michigan, 169–170
University of Rochester, xvii, 114
Uranus, as gas giant, 112–113
Urey, Harold, 103
USS *Theodore Roosevelt*, 68

Vallée, Jacques, 94

Variety, 62
Venus
   atmosphere of, 156–157
   brightness of, 29
   climate state of, 116
   Mariner 2 and, 33
   temperature on, 62
   terraforming and, 156
   as terrestrial world, 112
   visible from Earth, 2, 29
   water and, 118
Vernadsky, Vladimir, 135
volatiles, terraforming and, 155
von Danikein, Erich, 57
von Neuman, John, 180

warp drives (hyperdrives), possible interstellar travel and, 84
water
   comets and origin on planets, 119, 155
   life on ocean moons and worlds, 106–108, 118–121
   as prerequisite for life, 120
West, Mike, 71
Whitaker, Bill, 68
Wonderful Life: The Burgess Shale and the Nature of History (Gould), 169
wormholes, possible interstellar travel and, 83
Wright, Jason, 11–12, 141–143

About the Author

ADAM FRANK is the Helen F. and Fred H. Gowen Professor in the department of Physics and Astronomy at the University of Rochester. Frank earned his doctorate in physics at the University of Washington in 1992 and held postdoctoral positions at Leiden University and the University of Minnesota before winning a Hubble Fellowship in 1995. He joined the faculty of the University of Rochester in 1996. For many years his research group developed advanced magneto-hydrodynamic supercomputer simulation tools to study the formation and evolution of stars. He then turned to studies of exoplanet atmospheric evolution and astrobiology. His group's work on the "Astrobiology of the Anthropocene" were some of the first studies to look at anthropogenic climate change in a broadly astronomical context.

In 2019 Frank became the Principal Investigator on NASA's first-ever grant to study planetary technosignatures. The grant, involving researchers from across the US, was renewed in 2021 and has produced a variety of studies articulating the detectability of signatures of exo-civilizations on distant worlds.

Along with his scientific work, Frank has been a self-described "evangelist of science"—working to communicate the beauty and power of science—throughout his career. He was the cofounder of National Public Radio's popular *13.7 Cosmos and Culture* blog, which ran from 2009 to 2018, where he was also a frequent online NPR commentator. He now co-leads the 13.8 weekly column on BigThink.com. His writing appears in the *Atlantic*, the *New York Times*, and other venues, and he provides commentary on science for CNN and NBC. Finally, he was the science consultant for Marvel's *Dr. Strange* (best week ever!).

Frank has received a number of awards for his scientific and outreach work. His last book, *Light of the Stars*, won the 2019 National Honors Society Best Book in Science. In 2020 he was given the American Physical Society's Joseph A. Burton Forum Award. In 2021 he was granted the Carl Sagan Medal for excellence in public communication by the American Astronomical Society.